THE GERM-FREE ANIMAL
IN RESEARCH

THE GERM-FREE ANIMAL

IN RESEARCH

Edited by

M. E. COATES

The National Institute for Research in Dairying
University of Reading, Shinfield
Reading, Berkshire, England

Associate Editors

H. A. Gordon

University of Kentucky
Lexington, Kentucky, U.S.A.

and

B. S. Wostmann

Lobund Laboratory
University of Notre Dame
Notre Dame, Indiana, U.S.A.

1968

ACADEMIC PRESS · London and New York

ACADEMIC PRESS INC. (LONDON) LTD.
Berkeley Square House
Berkeley Square
London, W.1.

U.S. Edition published by
ACADEMIC PRESS INC.
111 Fifth Avenue
New York, New York 10003

Library of Congress Catalog Card Number: 68-24698

253885

PRINTED IN GREAT BRITAIN BY
ROBERT MACLEHOSE & CO. LTD
GLASGOW

List of Contributors

HEINZ BAUER, *Georgetown University Schools of Medicine and Dentistry, Washington, D.C., U.S.A.*

D. W. VAN BEKKUM, *Radiobiological Institute of the Organization for Health Research T.N.O., Rijswijk, The Netherlands.*

M. E. COATES, *The National Institute for Research in Dairying, University of Reading, Shinfield, Reading, Berkshire, England.*

T. Z. CSÁKY, *Department of Pharmacology, University of Kentucky College of Medicine, Lexington, Kentucky, U.S.A.*

R. FULLER, *The National Institute for Research in Dairying, University of Reading, Shinfield, Reading, Berkshire, England.*

H. A. GORDON, *Department of Pharmacology, University of Kentucky College of Medicine, Lexington, Kentucky, U.S.A.*

T. F. KELLOGG, *Lobund Laboratory and Department of Microbiology, University of Notre Dame, Notre Dame, Indiana, U.S.A.*

JULIAN R. PLEASANTS, *Lobund Laboratory, University of Notre Dame, Notre Dame, Indiana, U.S.A.*

BANDARU S. REDDY, *Lobund Laboratory, University of Notre Dame, Notre Dame, Indiana, U.S.A.*

E. SACQUET, *Centre de Sélection des Animaux de Laboratoire, Gif-sur-Yvette, Seine et Oise, France.*

JEAN-CLAUDE SALOMON, *Institut de Récherches Scientifiques sur le Cancer, Villejuif (Seine), France.*

P. C. TREXLER, *Department of Pathology, Royal Veterinary College, University of London, London, England.*

C. K. WHITEHAIR, *Department of Pathology, Michigan State University, East Lansing, Michigan, U.S.A.*

BERNARD S. WOSTMANN, *Lobund Laboratory, University of Notre Dame, Notre Dame, Notre Dame, Indiana, U.S.A.*

The contributors to this book wish to pay tribute to the memory of
James A. Reyniers, who died on November 3rd, 1967. He was one of the
great pioneers of germ-free research and his example has been an
inspiration to all of us. Much of the research described in this volume
rests on the foundations which his work provided.

Foreword

It was about sixty years ago that Professor Nuttal from San Francisco visited my chief, Professor Tangl, in the Laboratory of General Pathology of the University of Budapest. On that occasion he told us about his experiments with guinea pigs which he delivered by caesarean section and of his attempts to rear them under sterile conditions. Besides its scientific merit, maybe I remember this event so well because San Francisco was destroyed by an earthquake in the same year (1906). We were all afraid for the professor who fortunately was later reported to be safe. Perhaps I also remember this occasion because at the time we were all excitedly discussing the ideas of Metchnikoff of the Pasteur Institute in Paris who, in a new book — more like a novel than a scientific text — proposed that ageing was the result of a continuous intoxication caused by intestinal bacteria. Maybe I subconsciously connected these two impressive men of that year when I wondered whether an animal growing up in sterile conditions, without bacteria in its body, would ever age?

These old dormant memories were awakened some years ago when I saw the laboratories at the Lobund Institute at the University of Notre Dame and realized what had been achieved there, and by other pioneers in Sweden, Japan and elsewhere, through enormous effort, technical knowledge and skill. I was working on gerontological problems at that time and it seemed obvious to ask the question, how long do germ-free animals live? We now have the answer, not for maximal age, because it is too soon to decide that yet, but at least for the 50% survival time. It appears that in this respect germ-free mice have the advantage over comparable conventional controls.

Metchnikoff never experimented with ageing germ-free animals and the dream he inspired did not crystallize because it is now known that germ-free animals do age. But surprising differences in morphological, nutritional, immunological and pharmacological characteristics have become apparent. Now, offering a fount of new knowledge, the germ-free animal promises to become a commonplace tool in biological research. This volume contains the story of how the techniques have evolved, as so often happens, from complicated to simplified systems. It also tells of the facts already found, of experiments in progress and, last but not least, of future prospects.

vii

An enormous stimulus has been given by this new technique to the study of the influence of different diets when no bacterial breakdown or synthesis in the alimentary tract can take place. Is survival possible in species such as the ruminants whose nutrition depends on the microbial digestion of foods for which no endogenous enzymes exist? What happens to a rat or mouse on a vitamin-poor diet without the contribution of vitamins synthesized by intestinal bacteria? The fact that the flora is a direct or indirect regulatory factor governing the muscle tone of the intestine is a completely new concept, never suspected before. Lack of this exogenous factor may seriously impair normal organ function. Studies on such extreme effects as the inordinately large caecum of germ-free rodents are beginning to throw some light on the pharmacological processes concerned in normal gut function.

A new territory has been opened in immunological research. The role of various organs in body defences is becoming more clearly understood. The lymphoid organs of the germ-free animal, unchallenged by exogenous antigens, offer an excellent medium for the study of antibody synthesis. Infection of germ-free animals with known microbes is already answering questions regarding antibody response to specific immunological stimulation.

When Nuttal spoke to us sixty years ago, nobody suspected that there were living organisms smaller than bacteria. It was during World War I that we first heard of bacteriophages. Now we are aware of many viruses that may reach the embryo across the placenta and which seemed, until the discovery of interferon opened a new line of attack, impossible to destroy in the living body. This illustrates the complexity of the next step in gnotobiotics: the rearing of virus-free animals. Before this has been safely achieved, news is already reaching us (perhaps, as in politics, even before it is true!) that viruses may be produced by molecules of the body tissues, a revival of the long-believed *"generatio equivoca"*. The techniques used in gnotobiotic research may give answers to such problems which could influence the basis of our whole biological thinking.

We now live in an age where men may choose completely new milieus for living and at any time we may hear of a manned landing on the moon. It seems immensely important that we should be circumspect in introducing life from this earth to the moon or the planets, lest we obliterate the unique evidence that may exist there on the age-old problem of how life began in lifeless surroundings and what factors directed its further evolution. Who can prophesy how far gnotobiotic techniques may help us in such seemingly fantastic researches?

We owe a great debt to those who laboured so hard in an unknown desert to give us the techniques and to observe the characteristics of germ-free life. They have opened the road to the future understanding of the relationship between animals, including man, and their habitual microflora.

Institut für Experimentelle Gerontologie, Basel   F. VERZÁR germ-free life. They have opened the road to the future understanding of the relationship between animals, including man, and their habitual microflora.

Institut für Experimentelle Gerontologie, Basel   F. VERZÁR

Preface

By the end of the NATO Advanced Study Institute held in August, 1965 at Elvetham Hall, it was generally agreed that a book on germ-free research was needed and that the meeting provided a good starting point.

The papers had been outstanding; discussions had been both stimulating and informative and had added much to our understanding of the subject. It was felt that this collected knowledge should be given wider dissemination and that there were still some aspects of germ-free research that had not been fully covered at the meeting.

Many participants gladly agreed to rewrite and expand their contributions so that information gleaned from discussion could be incorporated and duplication eliminated. At the same time, other specialists not able to attend the meeting were recruited to write chapters. We owe much to the enthusiasm of our good friend Professor F. Verzár, who encouraged us to prepare this book; we hope that the result will be found sufficiently comprehensive and coherent to satisfy the requirements of all workers either in the field or entering it.

We are deeply grateful to all our colleagues who have so generously contributed from their expert knowledge and experience, in spite of many other demands on their time; to the friends who have read parts of the manuscript and offered helpful comments and advice; to Professor S. K. Kon for translating some of the chapters; to Mr. B. F. Bone for preparing the index and for invaluable assistance with the proof-reading and to Mrs. A. D. Pulling and Mrs. M. C. Lloyd for patient secretarial help.

May, 1968

M. E. COATES
H. A. GORDON
B. S. WOSTMANN

Contents

Chapter 1
EQUIPMENT DESIGN AND MANAGEMENT

Chapter 2
THE ROUTINE MICROBIOLOGICAL CONTROL OF GERM-FREE ISOLATORS

Chapter 1

Equipment Design and Management
Part I General Technique of Maintaining Germ-free Animals

E. SACQUET

*Centre de Sélection des Animaux de Laboratoire,
Gif-sur-Yvette, (S. and O.), France*

I. Introduction

Those research workers who first attempted to raise animals in the absence of microbial contamination had to achieve marvels of ingenuity and endurance. The hand rearing of young mammals essential to the creation of breeding colonies was a fearsome task. The diagram (Fig. 1) of the apparatus of Nuttal and Thierfelder (1897) allows one to appreciate the complexity of the outfits used at that time. Fortunately these days have passed. As forecast by Pasteur (1885), "Ces genres de travaux se simplifièrent par leur développement même". The breeding of axenic animals is no longer a prowess. The progress of the last ten years has led to the development of simple and efficient techniques and it is possible now, anywhere in the world, to send or receive germ-free animals. It is proposed to describe in this article the essential features which a research worker wishing to produce, and particularly to use, germ-free animals must know in order to establish a laboratory.

A. PRELIMINARY CONSIDERATIONS UNDERLYING THE ESTABLISHMENT OF A LABORATORY FOR GERM-FREE ANIMALS

An investigator intending to carry out germ-free work must define its objectives before thinking about the means of execution. This is a point often insufficiently considered. There are now so many ways available to choose from that the following questions deserve consideration. Is it really

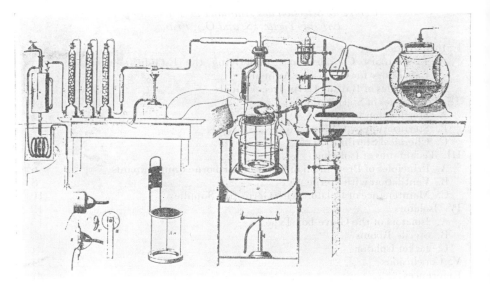

FIG. 1 Apparatus of Nuttal and Thierfelder (1897).

necessary to use axenic animals to carry out the proposed experiments? Is it necessary for the purpose to have a special set-up or would it be more convenient to carry out the work in an established laboratory? Is it necessary to breed the experimental animals or is it simpler to obtain them from a breeding centre? What type of isolator should one choose? How may the experimental material be adapted to conditions of work in a germ-free medium?

These questions, and many others, exercise the mind of the research worker. He is further perplexed when he considers the diversity of techniques open to him. There are many designs of apparatus intended for maintaining animals free from microbial contamination: Reyniers,

Gustafsson, Trexler, Phillips, Lev, Paterson and Cook have devised equipment of the glove-box type, that is, containers in which manipulations are carried out by a worker remaining outside the sterile area. Only the arms and hands enter the enclosed chamber by means of gloves forming an integral part of the wall of the isolator. Reyniers, Trexler, and De Somer and Eyssen with a group of engineers of Louvain (Belgium) have designed and built sterile rooms, which the worker enters wearing a diving suit which protects him from the chemical agent used to sterilize the outside of the suit. Midway between these two extremes, the glove-box and the sterile room, there is an intermediate type designed by Trexler known as a "jacket isolator". The numerous means of sterilizing these outfits and all that has to be introduced into them include dry heat, moist heat, radiation and chemical agents — peracetic acid, ethylene oxide, β-propiolactone, formaldehyde, etc.

The next few pages are intended to help the worker in choosing from the possibilities open to him. This article does not pretend to give a definite single answer or to determine at each stage what should be the equipment of a laboratory intended for the production or experimental use of gnotobiotic animals. Depending on the country, the scientific approach and the individual, the techniques will vary so as best to achieve their goal. For this reason it seems better to point out what particular advantages each offers to the user and the reasons underlying their invention.

B. PRINCIPLES OF RAISING GERM-FREE ANIMALS

To form a colony of germ-free animals it is necessary (i) to obtain animals devoid of any microorganisms and (ii) to maintain these animals protected from microbial contamination while supplying them with all they need to live and, if possible, to reproduce.

Since the world in which we live is contaminated by innumerable microorganisms, in order to protect a germ-free animal from microbial contamination it is necessary to isolate it by means of an enclosure (referred to as an isolator) which has previously been freed from microorganisms by sterilization and is capable thereafter of preventing the passage of any living microorganism. To achieve this object, efficient means of sterilization of the isolator are essential and the mechanical barrier must also be effective against microbial penetration. The efficacy must be absolute, as the penetration of a single microbial cell may cause contamination of the whole outfit. For this reason certain essential principles will be cited, known to any microbiologist, yet not always sufficiently rigorously applied.

II. Techniques of Sterilization

The containers or the material introduced into them may be sterilized in two ways: (*a*) by physical means such as heat or gamma irradiation; (*b*) by chemical agents.

A. STERILIZATION BY HEAT

Sterilization by steam pressure is mainly used. The most resistant microorganisms are killed by autoclaving for 15 min at 121°C (Perkins, 1957). Some, within the materials in which they are embedded, show a greater resistance. It is therefore advisable, when possible, to pre-sterilize the material, that is to submit it to a preliminary sterilization before the final sterilization in the isolators. It should be borne in mind that in autoclaving it is essential not only that the temperature be held at 121° C for 15 min but also that contact with steam vapour be maintained at 121° C during these 15 min. When only heat operates, as in the Pasteur oven, a temperature of 180° C for 30 min is needed to destroy the most resistant microorganisms. It is thus essential, when moist heat is applied, that steam should penetrate the material to be sterilized so as to be in contact with microorganisms that are to be destroyed. For this reason the construction of the object to be sterilized must be such as to allow access of steam. This point must be kept in mind in the manufacture of isolators and in the treatment of materials placed in them or introduced after sterilization. Air is an efficient barrier to the penetration of steam and tends to gather in the upper parts of the isolator. This air must be completely extracted, which requires a pump able to achieve a sufficiently high vacuum and with an adequate capacity relative to the volume of the isolator. A simple and efficient procedure consists in drawing a vacuum of about 65 cm of mercury and then introducing steam into the isolator without stopping the pump; in this way steam does not push the residual air into the crevices and cavities but helps to expel it. Finally, it is necessary that all parts of the equipment effectively reach the temperature of 121° C for the required time. Thus not only the volume but the thermal capacity of the object to be sterilized and also the output and temperature of the steam must be considered.

B. STERILIZATION BY GAMMA-RAYS

This technique is mainly used in England to sterilize food and surgical equipment. Irradiation is carried out in specialized establishments and those concerned with germ-free animals have only to send the necessary materials in sealed containers and establish that the dose of radiation applied suffices for sterilization with the least damage to the irradiated substance.

C. CHEMICAL STERILIZATION

A chemical agent must satisfy three essential conditions. (i) It must be capable of destroying all microorganisms at the concentration applied. (ii) It must remain in contact with the infective agent long enough to achieve destruction. (iii) It must then lend itself to removal from the isolator without leaving residues noxious to the germ-free animal or to the material contained in the isolator.

Among innumerable antiseptic agents only a few are capable of destroying not only resistant viruses and vegetative forms of bacteria but also spores. In fact, only formaldehyde, peracetic acid and ethylene oxide are used to sterilize isolators. Even so, it must be pointed out that we have no information on the efficacy of peracetic acid or of ethylene oxide with thick-capsuled eggs of parasites (e.g. oocysts of rabbit coccidia, eggs of *Hymenolepis*). It is known, nevertheless, that formaldehyde in solution is unable to destroy oocysts of rabbit coccidia, although they are destroyed by heat.

There must be contact between the germicidal agents and the microorganism it is to destroy. The powers of penetration of these various agents differ considerably. Ethylene oxide is reputed to have outstanding powers of penetration to the extent that it is difficult to contain it in enclosures believed to be air-tight. On the other hand, the penetrating power of formaldehyde or peracetic acid is much less and, when these antiseptics are used, it is essential to remove obstacles that might prevent their contact with the microorganism. A film of fatty material suffices to prevent the action of peracetic acid and it is therefore necessary to clean with detergents the surfaces it has to sterilize. An excellent procedure is to add to the solution of detergent 0·2–0·3% of peracetic acid, which disperses the fatty film and allows the bacteria contained in it to be killed. The penetrating power of formaldehyde is very low and it is essential that this antiseptic agent be used with enough steam which, in condensing, deposits the formaldehyde in the crevices of the enclosure.

Contact must be maintained for a sufficient time. Peracetic acid acts most rapidly; an 0·1% aqueous solution destroys in a few seconds spores immersed in it and the vapour rising from a 2% solution destroys spores in 20 min. To obtain the same result with formaldehyde vapour, spores must be exposed for at least 24 hr when 20 ml of 40% formaldehyde solution are used per m^3. Similarly, ethylene oxide is sluggish and requires several hours. Finally, it must be remembered that the activity of formaldehyde and of ethylene oxide varies with temperature.

It should also be pointed out that these various chemicals have some defects which must be taken into account. All three irritate the mucous membranes and some may be carcinogenic (Reyniers *et al.*, 1964). It is thus necessary to protect the worker and also the experimental animal by removing the vapours with a suitable mechanism. Peracetic acid is most corrosive; only stainless steel of the quality described as "acid resistant", pure aluminium (and not its alloys), certain plastics and Pyrex glass are completely resistant to a concentrated solution. The corrosive action is less with the 2% solution used for sterilization; polyvinyl chloride, certain synthetic rubber materials (neoprene, but not chlorobutyl) may be used under these conditions. On the other hand, ethylene oxide (Phillips, 1958) has no corrosive action and can be used to sterilize delicate apparatus. This is true only for ethylene oxide gas; in liquid form the product is corrosive and it is therefore essential that no condensation should take place during sterilization. Ethylene oxide forms an explosive mixture with atmospheric oxygen and should only be used mixed with carbon dioxide. Peracetic acid separates at a temperature of $-25°$ C from its aqueous solution and becomes explosive, hence it should not be kept at such a low temperature.

Some of these germicidal agents leave residues. Peracetic acid decomposes spontaneously into hydrogen peroxide and acetic acid, an innocuous product. Formaldehyde in solution leaves deposits of solid polymers of formaldehyde which are difficult to remove; it is necessary to ventilate for several days after treatment or to neutralize these residues with ammonia. Ethylene oxide tends particularly to form with various products (e.g. cellulose, some plastics) complexes which are difficult to break down and which remain irritant for a long time.

Thus the worker using these germicidal agents must have precise knowledge of those he applies. The agents have great advantages in that they enable thermolabile substances to be sterilized; not being used at increased pressure they can be applied in thin-walled enclosures of plastic; some, e.g. peracetic acid, are very cheap. But they vary in their properties, some of which are unfavourable, however, and have to be used judiciously. It is easy to understand that, in view of this diversity of characteristics, the opinions of research workers on the use of these agents vary considerably.

III. Techniques of Isolation

A. PRINCIPLES OF PREVENTION OF ENTRY OF AIR-BORNE CONTAMINANTS

The primary function of the wall of the isolator is to prevent the entry of microorganisms from the outside. Ideally this wall should, therefore, be fully air-tight and offer no openings for entry of bacteria or viruses. In fact, experience proves that many isolators, although not possessing this perfect air-tightness, are nevertheless able to prevent microbial entry provided the air pressure inside them is always slightly positive in relation to that outside. Under these conditions, even if there is a break in the wall of the isolator, the air movement is from inside towards the outside thus preventing the entry of microorganisms. Whatever the type of isolator it is thus necessary to arrange that at all times it remains in positive pressure in relation to the exterior. In the final analysis one can conceive an isolator as a volume of sterile air which, unfortunately, can only persist with a wrapping.

It is obvious that these considerations apply only when the sterile isolator is to be protected against contamination from the outside. The

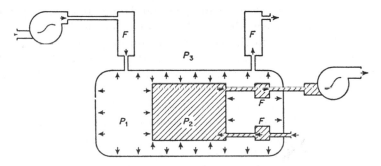

Fig. 2 Arrangement for ventilating an isolator while preventing its expired air from polluting the atmosphere. F, air filter; P, air pressure. (In this arrangement $P_1 > P_3 > P_2$, consequently the isolator, at P_2, cannot pollute the surrounding zone at P_1; the external atmosphere, at P_3, cannot penetrate the zone at P_1 and the isolator is thus protected against contamination.)

reverse problem, that of preventing a contaminated enclosure from polluting the air outside, is solved by the opposite procedure, ventilation by reduced pressure.

The most difficult problem, with which one is at times confronted, is that of preventing at the same time the penetration of air from the outside into the isolator and of any escape of air from the isolator to the outside. For example, if one wishes to carry out experiments on the germ-free animal with radio-nuclides, it is necessary on the one hand not to contaminate the animal and, therefore, to place it in a system of positive pressure in relation to the outside. On the other hand, to avoid pollution of the outside air by the expired air containing radioactive CO_2, the animal must be placed under negative pressure in relation to the outside.

In this situation it is necessary to use fully air-tight enclosures or, more easily, to place the ventilated isolator under negative pressure inside another enclosure under positive pressure in relation to the outside (Fig. 2).

B. VENTILATION WITH STERILE AIR

In view of the need to maintain the sterile enclosure under positive pressure it is obvious that isolators must be ventilated by outside compressed air forced into the isolator after passing through equipment that ensures sterility. Two procedures used to sterilize air introduced into the isolator are filtration and incineration. Filtration is more generally used. The filtering medium usually consists of four layers of special glass wool such as FG50 of the American Air Filter Co. Inc., Louisville, Kentucky, U.S.A. This material has the advantage of not being attacked by peracetic acid. There are also paper filters at least as efficient as FG50 but which should not be exposed to the corrosive action of peracetic acid. In the incineration procedure used by Gustafsson (1959), air travels through an apparatus where it is at the same time filtered and heated for a few seconds to 300° C.

Both procedures have proved their efficacy with bacteria and larger microorganisms. The incineration procedure ensures the destruction of viruses. Opinions are divided on whether filters are equally effective. Most research workers believe that viruses do not exist in the air as free particles a few millimicrons in size but as particles linked with much larger organic particles and that in consequence the viruses are retained with the larger bodies which carry them. Holmes (personal communication) has sought to confirm these theoretical considerations by passing aerosols of phages through filters under test and thence through cultures of sensitive bacteria; no phage could be demonstrated in the filtrates. Mattingly (1962) found glass-wool filters efficient in removing Polio II virus aerosols from air. It may seem bold to apply these findings to all laboratory conditions but filtration is a procedure so much simpler and less exacting, particularly when large volumes of air have to be handled, that it is the method of choice. Moreover, the results with it have so far been most favourable (Parker *et al.*, 1965) because there do not seem to be any reports in the literature of the contamination of germ-free objects by viruses having passed through isolator filters.

Whatever the procedure it is advisable to pre-treat the air pumped into the sterilizing equipment. The air is generally pre-filtered, which avoids clogging of the filters in the apparatus and increases safety of filtration. It is further conditioned as to temperature and humidity. This is done for two reasons: first, to ensure conditions of ventilation suitable to the standards of hygiene required for the raising of the animals in the isolator and secondly, to avoid condensation in the filters in which event the filters completely lose their efficacy. The first condition is generally satisfied by a humidity of 50–60% at the isolator temperature, by a temperature for small rodents of between 22–24° C and an air change of from 5–9 or more volumes per hr. When it comes to the second condition it is essential not only to avoid condensation in the isolator filter but also to ensure that the passing air does not contain water micelles in suspension forming a dry mist, as the micelles stopped by the filter would accumulate in it and wet it.

There are different ways of ventilating an isolator. The isolators are sometimes placed in a room where air is conditioned and partly replaced by outside air; the air from this room is passed into the isolators either by means of separate blowers for each isolator or a central blower supplying the whole assembly. Another procedure uses air taken from outside the laboratory; this air is conditioned as to temperature and humidity and then forced under pressure into the tubing feeding the isolators. The first technique is safer than the second which requires very reliable instrumentation without which the air is likely to be too dry in winter and saturated in summer. It is generally easier to condition the air of a room rather than to condition the few cubic metres necessary to ventilate the isolators.

It is obviously not possible to set the norms of pressure in a ventilating system. The pressure varies considerably with the type of filter. For paper filters pressures of 2–3 cm water are used; pressures of 5 cm of water frequently suffice for glass wool filters of the FG50 Trexler type. One has to bear in mind, however, the important losses of pressure that occur in the tubing and also that filters vary much in resistance depending on the way the glass wool is packed and also on the possibility that certain air intakes may be left open. For this reason a pressure of from 30–50 cm water seems preferable.

The exit of air from the isolator does not present much of a problem. Generally the outlet is equipped with an arrangement preventing the entry of air into the isolator should the pressure drop. In the Gustafsson design the air leaving the apparatus passes again through the incinerator. Occasionally a filter is placed at the outlet. It is my view that such an arrangement should be used only when essential, for example when the isolators are experimentally contaminated with infected agents that are dangerous to the workers. The outlet filters tend, in fact, to become clogged by dust coming from the isolator and some animals, birds in particular, produce much detritus likely to clog the filters rapidly.

C. MAINTENANCE OF ISOLATORS WITH GERM-FREE SUPPLIES

The germ-free animal must be supplied with water, food and, on occasions, bedding; it is also frequently necessary to introduce experimental material into the isolators.

1. WATER

Water is generally sterilized by autoclaving. It is usually pre-sterilized, the length of sterilization being roughly calculated with reference to the surface: volume ratio so as to allow for the time needed to bring the water to the requisite temperature. The isolators of Reyniers and of Gustafsson are supplied with equipment allowing sterilization of from 20 to 40 l of water.

Other containers are also used, for example tins, preferably soldered, and glass bottles with special stoppers or simply plugged with cottonwool. It is necessary to ensure proper sterilization of the stopper; whatever its type, the whole closing surface must be exposed to steam. Cotton wool plugs satisfy this condition provided the exposed surface of the plug is flamed to a depth of about 1 cm just prior to passage through a peracetic acid lock, in order to ensure the destruction of organisms that may have penetrated through the surface fibres.

2. FOOD

Sterilization of food is a delicate operation as not only must living microorganisms be removed but the nutrients should not be destroyed below the contents prescribed. The methods used are briefly described here. The subject is discussed in more detail in Chapter 4.

(a) Sterilization by Gamma Radiation

This method is applicable only to dry products. The foods are placed in air-tight containers often in several sealed layers of plastic. An elegant procedure used in England (Coates et al., 1963) consists in evacuating the envelope containing the food. A compact parcel is thus obtained and, should there be even the smallest leak, tautness is relaxed and the package is rejected as faulty. With this procedure great care is needed in introducing the package aseptically into the isolator. The outside wrappings are torn off under the protection of an antiseptic agent and the sterile bag is introduced into the isolator through the lock sterilized with peracetic acid or by immersion in an antiseptic bath. There is no doubt as to the efficacy of gamma radiation for sterilizing foods (Gordon, 1959; Horton and Hickey, 1961). Moreover, with proper dosage irradiation causes less destruction than autoclaving (Luckey et al., 1955; Kraybill,

1955; Tsein and Johnson, 1959; Coates *et al.*, 1963). Certain thermolabile substances such as antibiotics can thus be sterilized with ease (Luckey, 1963). Paterson and Cook (unpublished) have used this technique to sterilize freeze-dried cow's colostrum for a simple formula for the feeding of the baby rabbit.

(b) *Sterilization by Autoclaving*

This traditional procedure is still widely used. It has the advantage of being entirely under the control of the worker who can vary simply and at will the severity of the treatment. All that has been said earlier about the technique of autoclaving also applies in this context, particularly those passages pertaining to the penetration with steam of the substance to be sterilized.

It is most important that the texture of the food be such as to allow penetration of steam. Certain foods, although they do not contain heat-resistant bacteria, are not effectively sterilized by this procedure. Some soya-bean meals can be quoted in this connection as they form an impervious dough with the remainder of the diet. Certain foods, which can be sterilized as granules or pellets, cannot be sterilized as a powder, since in this form they contain much more air which prevents the penetration of steam. The water content is also important and should be maintained between 15 and 30%. Water plays a part not only in the destruction of microorganisms but also in the destruction of vitamins. Zimmerman and Wostmann (1963) have shown that the moisture content of a food profoundly influences the survival of B vitamins. The margin between the temperature allowing complete destruction of micro-organisms and that leading to the destruction of nutrients essential to higher animals, particularly those involved in reproduction, is very narrow. Many research workers have found it difficult to get germ-free rats to breed. In my experience sterilization of diet L 356 (Luckey, 1963) at 15·5 lb for 20 min has always proved satisfactory, whereas the same diet sterilized at 17 lb for 25 min failed to support reproduction because the mother rats became incapable of nursing their young. Zimmerman and Wostmann (1963) suggest taking the degree of destruction of thiamine as a reference standard for destruction of other nutrients in the diet, thiamine being the most thermolabile of them.

(c) *Chemical Sterilization*

Ethylene oxide has been suggested for sterilizing food because of its great penetrating power. The procedure has now been generally aban-doned as too destructive. However, recently, Charles *et al.* (1965) have not subscribed to this view: they are of the opinion that, if condensation of ethylene oxide during sterilization is carefully avoided, destruction is very small and that by suitable techniques it is possible after sterilization to remove completely residues of ethylene oxide.

3. BEDDING

Woodwool, sawdust, maize cobs and the like are easily sterilized by autoclaving. They should be given, for reasons already stated, a pre-sterilizing treatment.

4. EXPERIMENTAL MATERIALS

Sterilization of experimental materials presents many problems. Full appreciation of the basic rules relevant to various types of sterilization and precise knowledge of the chemical properties of different germicidal agents are essential to success. It is often necessary to modify conventional experimental apparatus before it can be effectively sterilized: a glass-blowing shop and a workshop for metal and plastics are therefore very useful. With these facilities available, and a judicious choice between sterilization with dry heat, moist heat (autoclaving), gamma radiation, formaldehyde, peracetic acid or ethylene oxide, a satisfactory solution can generally be found. It follows that a laboratory should preferably have all these means of sterilization at its command.

IV. Isolators

It would be rash to pronounce on the worth of the various types of isolator invented since the heroic days of Nuttal and Thierfelder (1895, 1896, 1897), Schottelius (1899, 1902, 1908, 1913), Cohendy (1912) and Cohendy and Wollman (1914). It would be equally bold to single out for choice any of the outfits recently invented by Reyniers, Gustafsson or Trexler, to whom every worker interested in germ-free life owes a great debt of gratitude for their accomplishments. There is no doubt that in these various outfits animals can be raised germ-free and that their value is established and it would be frankly wrong to blame the characteristics of any one isolator for failures encountered in its use. Though there are at present no serious difficulties in raising animals germ-free, any technical error has to be paid for and thus the human element is of the greatest importance. The contamination of an isolator is always the result of a mistake or of the negligence of the worker. In the following brief survey of isolators now in use their other characteristics and their value as a research tool will be stressed rather than their ability to maintain animals germ-free. Their inventors have fully described the different isolators and for details the original publications should be consulted.

A. ISOLATORS OF THE GLOVE-BOX TYPE

The dimensions of these isolators, provided with one or two pairs of gloves, do not exceed 1·6 m in length and 0·9 m in diameter when they are cylindrical, or 0·6 m in width when they are polygonal. The number of animals they can contain is thus restricted to, at most, ten cages of rats or thirty cages of mice.

. If one attempts to breed animals in such isolators for many generations a high degree of inbreeding results. If the initial conventional animals are not inbred, the mishaps of inbreeding arise after a few generations involving, in particular, a marked decline in reproductive performance. For this reason animals from different isolators are sometimes crossed, or fresh blood is introduced by obtaining new animals from a conventional colony by aseptic surgical delivery. It is obvious that with both these methods there is the risk of altering the microbiological status of the animals through the spread of microbial infections difficult to detect. The tendency is, therefore, to use inbred lines of rats or mice known to reproduce satisfactorily and to maintain the colony by strict inbreeding. In this way exchange between isolators may be avoided and with it the risk of contamination with infective agents difficult to detect. Moreover, should contamination occur only a small part of the germ-free colony is lost and the damage can easily be put right by multiplication of the animals from an isolator that has remained sterile. This last advantage is due to the small size of the isolators, of which it is possible to keep a fair number even in a relatively small room.

It is not easy to postulate the optimal dimensions of isolators. Usually the worker rapidly finds that his isolators are too small and too few. The size is limited by the radius of accessibility of the gloves and the need to balance carefully the size of the colony and the means of supplying it with water, food and bedding. Moreover, since the isolators have to be moved, their weight and size is important. Thus, if the experiment requires large numbers of animals or much experimental material it is better to increase the number of isolators and to interconnect them so as to obtain the necessary space.

1. ISOLATORS STERILIZED WITH STEAM

(a) *The Reyniers Isolator*

These outfits, the result of the long experience of Professor J. A. Reyniers and of his collaborators, have been very fully described by the authors (Reyniers *et al.*, 1946; Reyniers, 1959). In essence their walls are sufficiently thick to withstand the pressure of steam at 121° C used for the initial sterilization. A small autoclave ("food clave") forms an integral part of the assembly and serves to introduce material into the isolator after the initial sterilization. Having had experience with this type of apparatus since 1960, I would like to make a few comments.

These isolators have been conceived to form an absolute barrier, i.e. they have no opening that would allow microbial contamination. It is, in fact, possible to make them completely air-tight by taping the points of attachment of tubing with polytetrafluoroethylene tape and it is quite easy to check the absolute air tightness of the system with halogen gas under

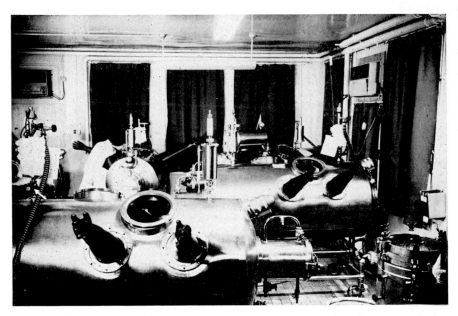

FIG. 3 The Reyniers isolator. The arrow indicates the valve that cuts off the filter from the rest of the apparatus and thus allows it to be dried at the end of the initial sterilization without the necessity of evacuating the whole isolator.

pressure. In use, however, leaks are likely to appear in two places, namely, at the jointings of the light-bulb portholes and at the point of attachment of the gloves. To avoid these faults it is sufficient to use synthetic rubber gaskets, that have a longer life, for the porthole joints and to use chlorobutyl gloves and to improve their method of attachment. Such gloves are not only more flexible than those made of neoprene but their resistance to mechanical or thermal damage is much greater.

Although the outfit is fully air-tight it is preferable to maintain enough positive air-pressure inside to avoid any possibility of the workers creating a negative pressure through their movements. For the same reason, it is useless, in my opinion, to draw a vacuum in the food clave after sterilization; the inside door of the food clave should be opened as soon as the steam pressure drops to zero. The small amount of steam that enters the apparatus can be rapidly removed through the air outlet trap. If a valve (Fig. 3) is inserted at the base of the air filter, it is also unnecessary to evacuate the isolator after the initial sterilization. When it is judged that the procedure has continued long enough to sterilize the filter, it is sufficient to close the valve and evacuate the filter in order to dry it. These simple modifications combine the advantage of full airtightness of the walls with the added safety of the principle of positive pressure. With these precautions the Reyniers isolator is a most reliable piece of apparatus and can be used for several years without resterilization. The isolator is self-contained, requires only a supply of air and steam and is independent of other autoclaves. It is the only one to be so independent and, as already said, it may be made absolutely air-tight. If the animals inside are infected with dangerous microorganisms a filter can be placed in the air outlet and the food clave can be used to sterilize infected bedding before removal.

The drawbacks of this type of isolator are its high cost, the need for a well-equipped workshop when it comes to the least modification and the smallness of the food clave which can, however, easily be made larger.

(b) *The Gustafsson Isolator*

This apparatus is designed for sterilization in an autoclave; as the steam pressure is thus equalized the thickness of the wall can be reduced and a large pane of glass can form the top. Since Gustafsson (1948) first described the apparatus, a side autoclave has been added for easy entry of water, food and bedding (Gustafsson, 1959). In this way the isolator has become a most valuable working outfit. The internal arrangement is such that space is completely utilized; in particular the high position of the autoclave allows the food, water and bedding supplies to be passed above the cages without their having to be moved. The large top window affords perfect visibility and it is at all times possible to introduce thermolabile substances through the built in germicidal bath. It should be pointed out, however, that because the germicide is a quaternary ammonium compound almost devoid of sporicidal power and gives off no sterilizing vapour, material introduced through such a bath must previously be very carefully cleaned and disinfected with more active agents (Gustafsson, 1959).

As all other isolators, except that of Reyniers, the apparatus requires a separate large autoclave, which must be considered in estimating the cost and planning the layout of the laboratory. With this reservation the

Gustafsson isolator offers the same advantages as that of Reyniers for a much lower price.

(c) *Other Isolators Using Steam as Sterilizing Agent*

Basing themselves on the Gustafsson method but using for walls plastics able to withstand autoclaving temperatures, Lev (1962) and Paterson and Cook (unpublished) have built very cheap isolators. They used nylon tubing of 60 cm diameter, produced commercially in England. These examples show how it is possible to use material of local production to build cheap apparatus.

2. ISOLATORS STERILIZED CHEMICALLY

(a) *The Trexler Isolator*

Trexler and Barry (1958) described an isolator made of thin (0·5–2 mm) polyvinyl sheeting, sterilized with a spray of a 2% aqueous solution of peracetic acid. Originally, supplies of water, food and bedding were steam-sterilized in a Reyniers isolator and were then transferred to the plastic isolator through a polyvinyl sleeve sterilized with the peracetic acid spray. Later, Trexler invented a simple set-up, described as a sterilization cylinder, which allowed the sterilization of supplies in ordinary autoclaves.

The method has recently achieved great success. By markedly reducing the cost of installation it has allowed the establishment of many germ-free centres, at first in the United States and later in less affluent countries. The success is due to the simplicity and efficacy of the method. The polyvinyl sheeting and the other materials constituting the isolator are cheap; further, polyvinyl sheeting is easily worked and welded. Thus, isolators of various sizes can easily be built, large pieces of apparatus can be introduced into them and the isolator can then be resealed and tubing and fitments of all sizes can be attached to the walls as they become needed in the course of an experiment. Furthermore, peracetic acid itself is also cheap.

Economy of time and labour is achieved by the use of sterilization cylinders for replenishing the isolators (see Fig. 4). Up to 10 cylinders can be sterilized in one autoclaving; one team of technicians may attach and sterilize the sleeves joining the cylinder to the isolator. As each operation takes only a few minutes, a series of five or six isolators can easily be serviced at the same time. Space can also be saved by placing the isolators in tiers, which is very easy with the necessary handling equipment.

The system is efficient. This efficiency stems from (a) the high value of peracetic acid as sterilizing agent, which acts rapidly on all the known microorganisms and kills spores; (b) the continuous maintenance of positive air pressure which keeps the polyvinyl envelope inflated; the withdrawal of the operator's hands from the gloves is counterbalanced by the corresponding collapse of the envelope. The positive air pressure prevents entry of contaminants in the case of a leak. In our experience it frequently happens that the floor of the isolator is punctured during an experiment but the damage does not result in contamination. The caps forming the doors of the lock are never air-tight but the positive pressure and the presence of peracetic acid ensure protection against contaminants.

Fig. 4 Cylinders for the sterilization of supplies to plastic isolators being introduced into an autoclave.

It goes without saying that the efficiency of the system depends on strict adherence to the proper use of germicides already mentioned. It is of particular importance that the lock through which the supplies are introduced be kept rigidly clean so as to ensure true contact between antiseptic and the wall. I prefer for this reason locks of rigid material, more easily cleaned, to locks made of pliable polyvinyl sheeting. Care must be taken during the spraying of peracetic acid into the lock that the pressure in it does not exceed the air pressure in the isolator, thereby avoiding the risk of penetration of organisms through the lock doors, which are never air-tight, before the germicide has acted.

Finally the technique of sterilization in cylinders must be thoroughly carried out. It is essential that the cylinders, apart of course from their glasswool filters, be air-tight and the welding should be checked with halogen gas. I have never discovered any leaks in the film used to cover the open end of the cylinder; it is an exceptionally pinhole-free material. All that is necessary is to attach it securely and protect it from damage. Finally, it is essential to remove air completely from the cylinders during the pre-vacuum stage of sterilization.

Of all these features the workers complain only of peracetic acid which is corrosive and irritates the skin and mucous membranes. It seems to me fairly easy to choose material that withstands the corrosive action, for instance, plastic cages with stainless steel lids and so on. The worker can also easily be protected by means of a corrosion-resistant, suitably placed fume exhaust. It is also possible to reduce the amount of fumes to which experimental animals are exposed by extracting the excess of acid deposited in the lock and by arranging the air outlet of the isolator near the lock so that the air current moves the acid fumes towards the exit. It is not, however, possible to remove the acid completely and thus the long-term effect of the fumes on the animals exposed to it is still open to doubt.

These considerations have led Kwa and Herbschleb (unpublished) to modify Trexler's apparatus so as to avoid the use of peracetic acid. The isolators are initially sterilized with ethylene oxide. They are supplied with a stainless steel air-tight door, which makes it possible to attach to the isolator a steam-sterilized lock. This lock joins the isolator to an equipment similar to the Reyniers outfit originally used by Trexler to supply his early plastic isolators. To avoid overheating of the polyvinyl film during steam sterilization of the lock, the area of the locked door is cooled by means of water circulating through a coil. The designers believe that they thus exclude unknown factors particularly undesirable in cancer research.

(b) *Rigid Plastic Isolators*

An isolator designed by Phillips *et al.* (1962) differs from the Trexler isolator in the nature of its walls, which are made of rigid acrylic sheet. It does not appear to be widely used, possibly because the acrylic sheet is more expensive and difficult to work than polyvinyl film. Furthermore, because of the rigid walls, a negative pressure can easily be created.

B. STERILE ROOMS

Various workers have attempted to raise germ-free animals not in small isolators accessible only to the hands of the worker but in sterile rooms of several cubic metres capacity which the worker himself can enter wearing a diving suit. These sterile rooms have several advantages over the glove-box type isolators. They allow the breeding of large numbers of individuals, inbreeding is reduced and the price of germ-free animals is cut down by reducing overheads. The number of animals it is possible to raise in a glove-box isolator declines sharply with the size of the animal. The rabbit already creates problems and pigs and ruminants cannot be kept in such an isolator beyond a certain size. These sterile rooms appear to be of potential value for the germ-free raising of large animals.

The sterile room of Reyniers (1956) was of cylindrical shape and had thick metal walls which allowed it to be sterilized by steam under pressure. The sterile rooms of the Rega Institute (de Dobbeleer, 1966) are of thin sheets of stainless steel and are chemically sterilized. Trexler (1959) covered the wall of a room with plastic resistant to peracetic acid which he used as germicide.

The diving suit for entering the room must, of course, be sterilized by chemical agents. Baths of formaldehyde solution, quaternary ammonium compounds or peracetic acid aerosol are used. The diving suit is the critical feature of the system. It must be so ventilated as to allow the worker to breathe and to cut down perspiration. Ventilation must be by

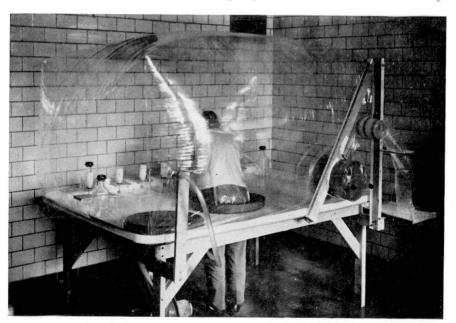

FIG. 5 Large jacket isolator. (Photograph by courtesy of P. C. Trexler.)

negative pressure since otherwise the smallest chink in the suit would contaminate the room and the negative pressure should be such as to prevent the worker from causing positive pressure anywhere by his movements. The technical problems of the diving suit do not seem to have

FIG. 6 Large jacket isolator with supply cylinder attached to the entry port.

been simply and satisfactorily solved, since the method of raising germ-free animals in germ-free rooms is still little used. Obviously, working in a diving suit creates various problems for the man wearing it. Moreover, should contamination occur many animals are lost. The method can only be worthwhile to the extent to which it is safe. For these reasons jacket isolators have become popular.

C. JACKET ISOLATOR

This type of apparatus (Trexler, 1961) is mid-way between the glove-box isolator and the sterile room. It is, in fact, the largest isolator that can be used without a diving suit. The worker reaches all parts of the enclosure by putting on a jacket which forms an integral part of the wall of the isolator and is attached to it through a circular opening in the floor (Fig. 5). In this way the worker is, in fact, outside the enclosure, as are the arms and hands of the worker in a glove-box outfit. Jackets suitable for prolonged use can easily be made and they can equally easily be ventilated by negative pressure relative to the isolator. Figure 6 shows a jacket isolator with a supply cylinder attached.

Such isolators have all the advantages of those of the glove-box type of Trexler but can be made much larger, for example, an isolator 2·5 m long, 2·0 m wide and 1·5 m high can accommodate a hundred and fifty mouse cages or seventy-five rat cages, or even more.

V. Conclusions

Technically it is not difficult to maintain higher animals rigidly germ-free. To achieve this object it is enough to follow the classic rules of sterilization and always to maintain in the enclosures in which the animals are kept positive air pressure which prevents the entry of microorganisms from the outside.

Many techniques, sterilizing agents and isolators have been described in recent years. All these techniques and equipment serve their purpose if the worker using them always keeps in mind the essential principles described here. The choice of a technique or material depends on the means of the worker and on the nature of the work he intends to carry out.

REFERENCES

Charles, R. T., Stevenson, D. E. and Walker, A. I. T. (1965). *Lab. Anim. Care* **15**, 321–324.

Coates, M. E., Fuller, R., Harrison, G. F., Lev, M. and Suffolk, S. F. (1963). *Br. J. Nutr.* **17**, 141.

Cohendy, M. (1912). *C.r. hebd. Séanc. Acad. Sci., Paris* **154**, 533–536.

Cohendy, M. and Wollman, E. (1914). *C.r. hebd. Séanc. Acad. Sci., Paris* **158**, 1283–1284.

de Dobbeleer, G. (1966). *Med. biol. Engineering* **4**, 227–260.

Gordon, L. E. (1959). *In* "The Becton, Dickinson Lectures on Sterilization 1957–1959", Seton Hall College of Medicine and Dentistry, Jersey City, N.J.

Gustafsson, B. E. (1948). *Acta path. microbiol. scand.* Suppl. 73, 1–130.

Gustafsson, B. E. (1959). *Ann. N.Y. Acad. Sci.* **78**, 17–28.

Horton, R. E. and Hickey, J. L. S. (1961). *Proc. Anim. Care Panel* **II**, 93–106.

Kraybill, H. F. (1955). *Nutr. Rev.* **13**, 193–195.

Lev, M. (1962). *J. appl. Bact.* **25**, 30–34.

Luckey, T. D. (1963). "Germ-free Life and Gnotobiology", p. 224, Academic Press, New York.

Luckey, T. D., Wagner, M., Reyniers, J. A. and Foster, F. L. (1955). *Fd Res.* **20**, 180–185.

Mattingly, S. F. (1962). *Am. J. Dis. Child.* **103**, 505–510.

Nuttal, G. H. F. and Thierfelder, H. (1895). *Hoppe-Seyler's Z. physiol. Chem.* **21**, 109–121.

Nuttal, G. H. F. and Thierfelder, H. (1896). *Hoppe-Seyler's Z. physiol. Chem.* **22**, 62–73.

Nuttal, G. H. F. and Thierfelder, H. (1897). *Hoppe-Seyler's Z. physiol. Chem.* **23**, 231–235.

Parker, J. C., Tennant, R. W., Ward, T. G. and Rowe, W. P. (1965). *J. natn. Cancer Inst.* **34**, 371–380.

Pasteur, L. (1895). *C.r. hebd. Séanc. Acad. Sci., Paris* **100**, 68.

Perkins, J. J. (1957). *In* "The Becton, Dickinson Lectures on Sterilization 1957–1959", Seton Hall College of Medicine and Dentistry, Jersey City, N.J.

Phillips, A. W., Newcomb, H. R., LaChapelle, R. and Balish, E. (1962). *Appl. Microbiol.* **10**, 224.

Phillips, C. R. (1958). *In* "The Becton, Dickinson Lectures on Sterilization 1957–1959", Seton Hall College of Medicine and Dentistry, Jersey City, N.J.

Reyniers, J. A. (1956). *Int. Congr. Biochem. III*, Brussels. 458–465.

Reyniers, J. A. (1959). *Ann. N.Y. Acad. Sci.* **78**, 47–79.

Reyniers, J. A., Trexler, P. C. and Ervin, R. F. (1946). Lobund Rep. No. 1. University of Notre Dame Press, Notre Dame, Indiana.

Reyniers, J. A., Sacksteder, M. R. and Ashburn, L. L. (1964). *J. natn. Cancer Inst.* **32**, 1045–1056.

Schottelius, M. (1899). *Arch. Hyg. Bakt.* **34**, 210–243.

Schottelius, M. (1902). *Arch. Hyg. Bakt.* **42**, 48–70.

Schottelius, M. (1908). *Arch. Hyg. Bakt.* **67**, 177–208.

Schottelius, M. (1913). *Arch. Hyg. Bakt.* **79**, 289–300.

Trexler, P. C. (1959). Proc. 2nd Symposium on Gnotobiotic Technology, University of Notre Dame Press, 121–125.

Trexler, P. C. (1961). *Bio-Medical Purview* **1**, 47–58.

Trexler, P. C. and Barry, E. D. (1958). *Lab. Anim. Care* **8**, 75–77.

Tsein, W. and Johnson, B. C. (1959). *J. Nutr.* **68**, 419–428.

Zimmerman, D. R. and Wostmann, B. S. (1963). *J. Nutr.* **79**, 318–322.

Part II Transport of Germ-free Animals and Current Developments in Equipment Design

P. C. TREXLER

The Charles River Breeding Laboratories, Inc., Wilmington, Massachusetts and Snyder Laboratories, New Philadelphia, Ohio, U.S.A.

I. Introduction

The gnotobiote as a laboratory animal should be available in sufficient numbers within a limited weight range or age-span to provide adequate groups of animals for study. Only a very large installation can produce sufficient animals for a variety of investigations, hence the need for transporting these animals from large breeding colonies to the user's laboratory.

It is clear from the foregoing discussion that the apparatus and methods now available for the maintenance of gnotobiotes provide sufficient security so that contamination can be avoided. The animals and other biological materials within the isolator can be observed and manipulated in a manner comparable to that of the open room. The principal limitation in the use of isolators involves the introduction and removal of supplies and experimental materials. Efforts are being made to improve sterile transfer procedures and to provide a more convenient and less expensive way of using gnotobiotes through disposable isolators.

II. The Transport of Gnotobiotes

Gnotobiotes can be transported from one laboratory to another provided they are placed in an isolator that is small enough to be moved readily and air is supplied through electrically or mechanically operated pumps (Gustafsson, 1959; Trexler, 1959a). If these isolators are to be transported by a common carrier they must be sturdy enough to withstand rough handling or be placed in a protective shipping box. The animals may be used at the receiving laboratory either within the isolator in which they were sent or they can be transferred to other equipment. Shipment of gnotobiotic animals involves all the hazards associated with the shipment of conventional laboratory animals, such as the detrimental effects of extreme temperatures or delays in transit.

A. LOBUND-TYPE SHIPPING ISOLATORS

A shipping isolator developed at Lobund is shown in Fig. 1. This unit is a modification of an isolator frequently used in the laboratory (Trexler and Reynolds, 1957). The animals are housed in a polyvinyl film chamber equipped with arm-length rubber gloves, inlet-outlet filters and an entry port. The unit is attached to a substantial base, over which a protective cover is placed. This cover is made of aluminium with glass wool insulation on the inside. Air is supplied to the unit by means of a battery-operated blower. Aircraft-type lead-acid batteries are used in order to reduce the hazard of spillage. Sealed batteries such as the nickel-cadmium type would be more suitable but they are considerably more expensive.

The isolator is sterilized with peracetic acid or ethylene oxide. If peracetic acid is used, the filters are usually sterilized in the autoclave. After the isolator has been thoroughly vented to remove the sterilant, the animals are transferred in disposable cages from rearing isolators. The cage tops are fastened securely and the cage tied to a floor sheet of stiff polyethylene by means of flexible ribbons. These may be cut from polyvinyl film, or any of the wide variety of nylon or polypropylene strap used to secure boxes and crates is also suitable. Additional supplies such as sterile food, bedding or water can be placed in an animal cage and secured to the floor sheet. Pelleted feed may be soaked with water so as to provide both water and food during the trip. A more satisfactory ration can be prepared by mixing water with a meal to which about 2% agar has been added. The dry ingredients must be mixed thoroughly and 70–80% water added to provide a firmly gelled mix. The mixture can be placed in a tray about one inch deep and autoclaved in a sterile drum. Care should be taken to sterilize the mixture thoroughly since the steam will not penetrate a mash as readily as pellets (cf p. 11). After the food has been autoclaved and cooled, it may be broken into chunks and placed in the cages on top of the bedding.

Several thousand shipments have been made both from research institutes and commercial colonies in this type of shipping isolator. The hazard of contamination is no greater during transport than in the laboratory. Animals may be stressed in extremely hot or cold weather in much the same way as with the shipments of conventional animals. The

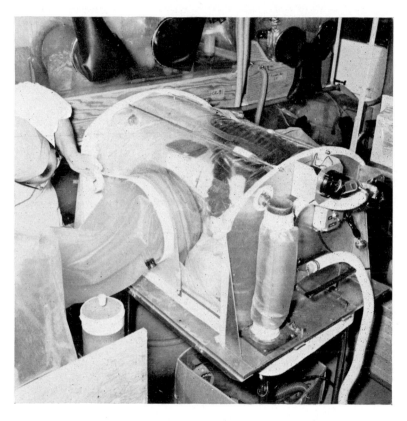

Fig. 1 Lobund-type shipping isolator. The isolator is shown connected to the rearing isolator prior to transfer of the animals. The battery holder is underneath the temporary room-air supply hose. The protective cover attaches to the base and is supported by the rigid end frames. (The Charles River Breeding Laboratories Inc.)

relationship between the battery size and power requirements of the blower will determine the time span through which air will be provided to the isolator. Twenty-four hours seems adequate for shipments within the United States and forty-eight hours for international shipments.

B. REYNIERS SHIPPING ISOLATORS

1. STEEL CYLINDER UNIT

Steel cylinders approximately 16 in. diameter and 24 in. long have been described by Reyniers (1957) for transporting animals between laboratories using the Reyniers system of steam-sterilized isolators. These units are attached to a special end flange on a Reyniers isolator for loading or unloading. The entire isolator with the shipping unit must be steam-sterilized. After cooling and drying the unit, animals may be transferred into it from another isolator. The door of the shipping unit is bolted shut and the unit, which bears sterilizing inlet and outlet filters, is ready for shipping. It must be attached to an air supply to provide ventilation. A special air-conditioned station wagon is used to transport the unit between laboratories. The receiving isolator is steam-sterilized after the shipping unit is attached.

This type of shipping isolator is heavy, since it must be capable of withstanding an internal pressure of at least 15 lb per in² for the initial sterilization. It is difficult to control the temperature in the animal cages during the steam sterilization of the receiving isolator. For this reason the arrangement was not considered practical and after a few initial trials was discarded.

2. ALUMINIUM-PAN SHIPPING UNIT

Reyniers and Sacksteder (1958) have described a transport unit made from a domestic aluminium roaster which can be attached to the steam-sterilized Reyniers isolator. This unit carries a single flat air-filter made from the same fibreglass filter material as used on rearing isolators. The entire unit is steam-sterilized within an autoclave or Reyniers isolator. After vacuum drying the unit is attached to a Reyniers isolator by means of a special connecting lock. The interior of the connecting lock is sterilized with germicide. The entire interior of the shipping isolator serves as the animal cage. Apparently this type of shipping isolator has not been used very much.

3. GLASS JAR SHIPPER

Reyniers and Sacksteder (1959) described a simple shipping container made from a glass jar of the type used for bottling fruit and vegetables. Fibreglass filter material of the type used for sterilizing air is placed in the lid. The entire jar and lid are sterilized by steam in the entry lock and passed into the interior of the isolator. The animals with bedding and food are placed in the jar and the lid tightened. The jar may then be passed out through a sterile lock or a germicidal trap. To reintroduce the jar into a sterile isolator, the surface of the jar must be sterilized with a germicide either within a sterile lock or bath. A surgical glove can be placed over the filter and around the jar lid to avoid contaminating the animals with the sterilant. If the jar lid is removed carefully so as not to disturb the surfaces concealed by the glove, the animals may be removed without the introduction of microbial contamination.

This device has been used successfully by several workers to send small numbers of animals from one laboratory to another or to remove animals from an isolator for short periods. Care must be used in reintroducing the jar into an isolator in order to avoid contamination. The rubber glove or other protective dam used over the filter must not be disturbed while in the sterile isolator.

C. THE BAKER SHIPPING ISOLATOR

Baker (1965) described a disposable shipping unit made from a four-ply fibreboard drum, 9 in. outside diameter, 10 in. in height. Several large holes are cut into the drum; these are covered with wire mesh and three layers of half-inch fibreglass filter medium. The filter medium is then covered with protective screening held in place by metal straps and the unit is steam-sterilized. The drum is reinforced with plastic tapes which are also used to hold the lid to the body of the drum. The body of the drum is attached to a plastic sleeve which is then attached to the entry port on an isolator so as to serve as a sterile lock. The lock is sterilized with a peracetic acid spray. The shipping unit is then introduced into the sterile isolator. The tape holding the lid in place is removed and the animals with bedding placed inside the drum. Food and water is provided by means of a porridge consisting of six parts water and four parts oats, sterilized with the shipping isolator. The lid is then taped in place and the shipping isolator removed. The process is repeated in reverse for transferring the animals into the isolator.

This shipper is inexpensive and can be disposed of at the receiving laboratory. Baker (1965) reported that over five hundred units had been used in the shipment of some 8500 mice. Only four losses occurred and these were due to damage to the shipping unit.

D. THE CHARLES RIVER SHIPPING ISOLATOR

Over the past four years, a lightweight disposable shipping isolator has been developed, based upon the use of a flexible plastic sleeve to contain the animal cages (Trexler, 1966). The animals are kept in cages in order to facilitate their introduction and removal from the shipping unit and to isolate small groups of animals according to size, sex or litter. The sleeve isolator also can serve as a holding chamber for animals, provided it is periodically attached to a larger isolator so that the cages can be serviced. A cardboard carton is used to protect the isolator during transit.

The shipping chamber consists of a tube of flexible plastic film, 12 in. in diameter and approximately 42 in. long. To provide clarity and

FIG. 2 The Charles River shipping isolator. The isolator is shown attached to the rearing isolator. Note the empty disposable cages with retaining straps in the shipper. (The Charles River Breeding Laboratories Inc.)

strength, pressed polished plasticized polyvinyl chloride approximately 0·02 in. thick is used. One end of the sleeve is closed by a filter consisting of two layers of FM004 filter medium, $\frac{1}{2}$ in. thick (Owens Corning Fiberglas Corporation, Toledo, Ohio). The surface of the filter medium is protected by a fibreglass scrim. The fibreglass and protective scrims are securely taped around a rigid support ring to provide a firm base for attachment to the sleeve. The upper filter is made in a similar fashion except that an oval rather than a circular support ring is used (Fig. 2). Two disposable cages with tops are placed within the sleeve together with a cardboard spacer approximately $1\frac{1}{2}$ in. high, in order to keep the upper filter away from the cage tops. This provides better ventilation and prevents the animals from damaging the filter. An oval shaped door is used to close the open end of the chamber. This door is made from a rigid plastic support ring, covered by a thin film taped in place, so that access to the cages may be had either by removing the door or rupturing the closing membrane.

The isolator is ordinarily sterilized by ethylene oxide within an autoclave equipped with a high vacuum pump. The door, described above, is placed loosely within the sleeve in order to assure that all surfaces are sterilized. The end of the sleeve is closed by another door which is taped in place.

After sterilization, the isolator is left in a well-ventilated place for at least 24 hr in order to ensure the removal of all ethylene oxide. The door, which has been lying loose in the sleeve, is manipulated into place and is secured with adhesive tape or a rubber band applied to the outside of the sleeve. The protective door, which has been used to close the end of the sleeve during sterilization, is removed, the open end of the sleeve sprayed with peracetic acid and then attached to the entry port (Fig. 2). After 20 min, the door is loosened and passed into the supply isolator together with the cages. The cages are loaded with bedding, food and the animals, the tops are strapped down and the cages returned to the shipping isolator. The door is replaced and taped securely, after which the shipper is removed (Fig. 3). The isolator is then placed in a protective cardboard carton for shipment (Fig. 4). When received, the process is reversed to introduce the animals to the receiving isolators.

Over five thousand shipments have been made with this type of isolator. Approximately forty mice or twenty rats can be sent in the current model. The size of the animals and the climatic conditions determine the numbers that can be sent without loss. However, the use of a microbiologically secure filter without forced ventilation makes it imperative to correlate the size and number of animals with the environmental conditions during transit. Ordinarily, delivery is made within 24 hr, although 48 hr has been used as the test period in design studies. The problems involved in using this isolator are much the same as those of shipping ordinary laboratory animals.

The current model effectively prevents contamination of the animals during transport. Contaminations that occurred were usually due to an error in technique in the receiving laboratory. These are often difficult to evaluate. Errors in the supplier's laboratory are more readily detected since ordinarily several shipments are made at the same time and some stock is left in the supply isolator for testing. Only two layers of filter medium are used on the shipping isolator while three or four are used on the regular laboratory isolators. The reduced thickness of filter does not present a contamination hazard provided air is not forced through it. A colony of gnotobiotic mice has been maintained for several years without

FIG. 5 The Charles River shipper. The cages have been loaded with animals, food and bedding, the end door taped in place and the unit disconnected from the rearing isolator.

contamination traceable to the use of these filters (M. M. Rabstein, personal communication). Four to five of these flat filters were sealed in the walls of a 2 ft × 2 ft × 4 ft flexible film chamber. No forced ventilation was used and the mice appeared to do very well in spite of 90% relative humidity within the chamber.

The shape of the chamber of the shipping isolator has been modified to provide means for withdrawing specimens for microbiological examination before the animals are passed to a receiving isolator. The chamber diameter was increased to 18 in. and a pair of wrist-length gloves were attached to short sleeves sealed on the walls of the chamber. Specimens

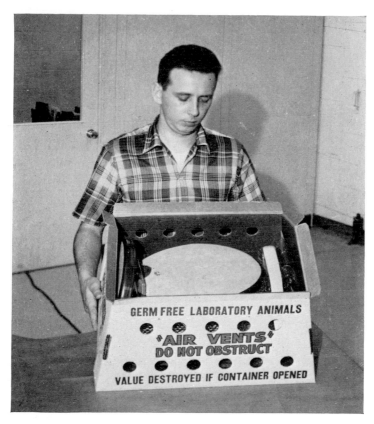

FIG. 4 The Charles River shipper. The isolator has been placed in the protective carton. A top will be strapped on the box before it leaves the laboratory.

were placed in a plastic pipe nipple, 1 in. diameter and 4 in. long, sealed in the wall of the chamber. The opened ends were closed with rubber stoppers so that the specimen could be removed without introducing contamination. This modification can be carried further so as to provide an isolator that can be used for shipment as well as maintenance of animals in the laboratory.

III. Current Developments in Equipment Design

At present there appears to be limited interest in the further development of apparatus for use with gnotobiotic animals. Apparently apparatus and methods available are adequate for many uses although there is need for larger numbers of animals, other species, particularly the larger domestic animals, and greater environmental control. These requirements justify further development of apparatus and methods.

A. STERILIZING FILTERS FOR AIR SUPPLY

Air filters consisting of 3 or 4 layers of $\frac{1}{2}$ in. FM004 glass fibre filter media (Owens Corning Fiberglas Corporation, Toledo, Ohio) wrapped around a cylindrical support (Trexler and Reynolds, 1957) have been widely used for sterilizing air for gnotobiotic animals. These cylindrical supports have been made from heavy welded steel. Recently, light-weight less expensive supports have been introduced consisting of a length of perforated polypropylene tube, the ends of which are closed with a polypropylene bottle split in the equatorial plane (Snyder Laboratories, New Philadelphia, Ohio). These filters are light enough to be supported by the walls or ceiling of the isolator chamber. They may be used either external to or within the chamber in an out-of-the-way corner. Filters can be made in any length up to 6 ft. They provide a low velocity inlet for the air so that better ventilation patterns can be maintained within the isolator.

Flat filters that provide ventilation without the use of fans or blowers have been described previously for the shipping isolators. They have been used successfully for the production of animals (Rabstein, 1964). However environmental conditions are of importance in many studies and control groups of animals associated with microorganisms but maintained in isolators are necessary. Under these circumstances it is necessary to maintain humidity and other components of the atmosphere equivalent to those of the open animal room. These can be maintained more readily within a closed chamber by means of forced ventilation.

A variety of filters has been developed for controlling particulate airborne contamination in the micron and sub-micron size range. These are more subject to damage particularly during sterilization in the autoclave than the type mentioned above. In a recent survey (Kretz and Ernst, 1966) sixty-eight out of seventy commercially-made filters of this type were found to permit the passage of bacterial spores. Filters of this type have less resistance and are more compact than comparable filters made from glass filter material but whether or not they can provide the security required for maintaining gnotobiotic animals remains to be demonstrated.

B. DISPOSABLE ISOLATORS

By definition, a disposable item is one with an initial cost so low that cleaning and repair required for re-use are not justified. The initial cost of an isolator for use with gnotobiotic animals is usually not a major item. The investment in time and effort required to obtain and use the biological material is usually much greater. For this reason an isolator to be disposable must, in addition to having a low initial cost, be reliable so as not to risk

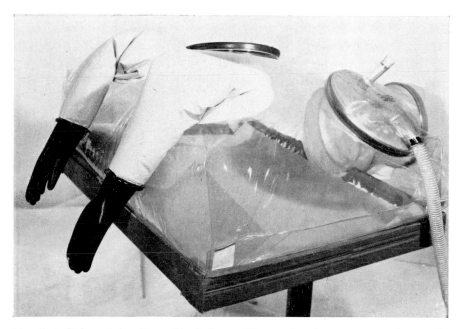

FIG. 5 A lightweight disposable isolator. The entire floor area can be easily reached through the gloves since they are attached to the roof of the chamber. The flat filters are made with 3 layers of filter medium in order to sterilize the ventilating air stream. The sterile lock is shown opposite the gloves. (Snyder Laboratories, Inc.)

contamination, be easy to use so as not to increase labour cost and be capable of maintaining an adequate environment to facilitate the design of experiments.

A promising design for such an isolator is shown in Fig. 5. The chamber is made from a flexible film tube, sealed at each end with a single straight seal. An entry port similar to the Charles River shipper is used for the introduction and removal of animals and supplies. The animals are handled in cages so that they can be maintained in much the same manner as conventional animals. The isolator is light-weight and can be readily stacked in racks to conserve space. This shape of chamber is easier to use than the larger chambers in which the gloves are attached to the front panel. The plastic film on the models tested was laminated or press polished in order to eliminate the danger of pinholes in the material. It seems likely that quantity production will make it possible to reduce the cost so that the isolator could be considered disposable. Gnotobiotic animals have been maintained in these isolators for a year without difficulty. The unit appears as reliable as other flexible film isolators.

C. TRANSFER PROCEDURES

Transfers in and out of flexible film isolators are ordinarily made by means of a sterile lock or antechamber in which a 2% peracetic solution is used as a sterilant. This procedure has been adequate for the maintenance of sterility. However, it does place a stress upon both the animals in the isolator and the attendants, due to the irritating effects of the vapors on skin and mucous membranes. Rats and mice are noticeably irritated by the fumes whenever supplies are introduced into the isolators, whilst Heneghan and Gates (1966) reported on the effects of the sterilant on gnotobiotic dogs. The peracetic acid within the sterile lock can be readily removed by installing a simple outlet trap or filter at the lowest point so that liquid accumulating in the sleeve can be drained away. The cap on the inside opening of the lock can be partially opened so that air from within the isolator will flush through the sterile lock. After sterilization, peracetic acid can be neutralized with an alkali such as aqueous sodium bicarbonate solution. This solution can be kept in the isolator and nebulized into the chamber before transfers are made. However, care should be taken to remove the residue before attempting resterilization.

Connections can be made between sterile isolators and/or supply drums by attaching two flexible membranes together and then cutting through the area of attachment. Microbes on the non-sterile side of the membrane can be immobilized by cement or fused plastic (Trexler, 1959b), or they may be inactivated by means of a liquid germicide. Pilgrim and Thompson (1963) described a transfer system in which two isolators are connected together through steam sterilizable membranes that terminate the distal end of rubber sleeves attached to the respective isolators. The external surfaces of these two membranes are swabbed with tincture of iodine, they are then clamped together and a passageway is cut. A somewhat similar technique has been used to introduce instruments into a sterile isolator for human surgery (Levenson et al., 1964).

REFERENCES

Baker, D. E. J. (1965). *Lab. Anim. Care* **15**, 432–439.

Gustafsson, B. E. (1959). *Ann. N.Y. Acad. Sci.* **78**, 17–28.

Heneghan, J. B. and Gates, D. F. (1966). *Lab. Anim. Care* **16**, 96–104.

Kretz, A. P. and Ernst, R. R. (1966). *Contamination Control* **5**, 18–26.

Levenson, S. M., Trexler, P. C., LaConte, M. and Pulaski, E. J. (1964). *Am. J. Surg.* **107**, 710–722.

Pilgrim, H. I. and Thompson, D. B. (1963). *Lab. Anim. Care* **13**, 602–608.

Rabstein, M. M. (1964). 'Gnotobiotic Symposium and Workshop.' Michigan State University.

Reyniers, J. A. (1957). *Proc. Anim. Care Panel* **7**, 9–29.

Reyniers, J. A. and Sacksteder, M. R. (1958). *Appl. Microbiol.* **6**, 146–152.

Reyniers, J. A. and Sacksteder, M. R. (1959). *Proc. Anim. Care Panel* **9**, 97–118.

Trexler, P. C. (1959a). *Ann. N.Y. Acad. Sci.* **78**, 29–36.

Trexler, P. C. (1959b). *Proc. Anim. Care Panel* **9**, 119–125.

Trexler, P. C. (1966). *U.S. Pat.* No. 3, 238, 922.

Trexler, P. C. and Reynolds, L. I. (1957). *Appl. Microbiol.* **5**, 406–412.

Chapter 2

The Routine Microbiological Control of Germ-free Isolators

R. FULLER

National Institute for Research in Dairying,
(University of Reading), Shinfield, Reading, England

I. General Considerations

It is self-evident that for a germ-free system to be of any value it should be known to be germ-free. For this reason a reliable system of sterility testing is of paramount importance. Procedures have been described and evaluated previously by Wagner (1959a) and sterility tests in current use are based on his recommendations.

There is a limited amount of information available on the subject of sterility testing of germ-free isolators. This deficiency is partly due to the fact that descriptions of sterility breakdowns are not relevant to the project as it is finally reported and partly to a reluctance to air in public what could be construed as a personal failure. The following discussion is an attempt to restate the general principles and point out some of the limitations and pitfalls of the microbiological procedures in routine use, as illustrated by experiences in this and other laboratories. As such, it does not embrace a detailed description of tests for viruses or physical tests for leaks.

A. THE PURPOSE OF STERILITY TESTING

The aim of sterility testing in the context of gnotobiotics is to establish that the experimental animal is not contaminated with other viable organisms growing in association with it, i.e. growing in the animal's tissues or in its immediate environment. It is necessary to qualify the definition in this way because it is impossible to establish beyond all doubt the complete absence of viable organisms without going to the ridiculous lengths of culturing all the material in the isolator. Nor is it necessary for the isolator to be germ-free in the strictest sense since it is quite possible for some organisms to gain access to the sterile area without being able to multiply and cause detectable contamination. In cases such as this, the more diligent the search, the more likely is contamination to be detected.

As with all types of sterility testing, the compromise which is arrived at is the examination of a sample and an assumption that it is representative of the environment inside the isolator. This procedure enables one to state that within the limits of the technique used no microorganisms of the type tested for are growing in association with the experimental animal. An extensive test procedure such as that outlined by Wagner (1959a) would enable one to make such a statement with regard to arthropods, worms, protozoa, bacteria, fungi, rickettsiae and viruses.

Wagner (1959a) has pointed out the limitations of attempts to establish the germ-free status of an animal. The basic weakness is that germ-freeness is a negative characteristic, i.e. it is based on the failure to recover viable organisms from the samples. This in turn is dependent on the techniques available and it is quite possible that at any given moment in time the techniques used will be inadequate. For this reason it is impossible to know for certain whether past reports of germ-free animals are valid. The inadequacy of early techniques and the gradual increase in complexity of sterility testing procedures is amply illustrated by Luckey (1963) in his historical survey of this topic.

It follows, therefore, that while a negative test result indicates that the animals are free from those organisms tested for, future development may show that the tests were far from adequate. As Luckey (1963) points out, "Until we can define life we cannot hope to devise tests to detect all variations of life . . .".

B. THE FATE OF ORGANISMS GAINING ENTRANCE TO GERM-FREE ISOLATORS

It is probable that more organisms gain access to the isolator than are ever detected by the sterility tests. The outcome of an organism getting inside the isolator can be as follows. A: The organism dies of starvation as in the case of a vegetative cell coming to rest on a stainless steel surface. B: It remains viable but does not multiply because it has no nutrients available or because of the presence of bacteriostatic levels of inhibitors. This latter complication may result from the coating of spores with residues of the $HgCl_2$ used for sterilizing eggs. An unusual example of this type of localized contamination is reported by R. Kenworthy (personal communication). During the course of hatching eggs in a germ-free isolator he noticed that one of the eggs turned almost black. After the other eggs had hatched he removed the black egg and on examination found it to be heavily contaminated with a coccus and a Gram negative rod. In spite of this the isolator remained germ-free for six weeks. C: The organism multiplies on the excreta, diet or water but does not colonize the animal. D: It colonizes the animal without growing in the environment. E: It colonizes the animal and grows in the environment.

Situations A and B constitute no risk to the experiment but in a strict sense the isolator is no longer germ-free. Situations C, D, and E will all give rise to positive sterility tests but have a varying effect upon the experimental results. If the organism is growing in the environment but not in the animal the worst that can occur is that the animal ingests large numbers of organisms which have multiplied in the water and diet. These organisms fail to colonize the gut because conditions are unsuitable in that site for a variety of reasons. Gibbons et al. (1964) experienced difficulty in establishing certain organisms in germ-free mice. Although most of their organisms colonized the gut when administered as poly-contaminants, Treponema microdentium and Bacteroides melaninogenicus did not. It was suggested that since T. microdentium required a relatively narrow range of oxidation-reduction potential for initiation of growth (Socransky et al., 1964) the conditions in the alimentary tract of gnoto-biotic mice were unsuitable. B. melaninogenicus, it was found, was inhibited by alfalfa in the diet. When this was omitted the organism be-came established in the gut. Other factors may make establishment difficult in the gut. For example, long chain fatty acids of host origin have been shown to reach inhibitory levels in the alimentary tract of conventional pigs (Fuller and Moore, 1967). Another example of this type of breakdown is one which occurs in a large proportion of our contaminations with germ-free chickens. Most of these contaminations are due to Bacillus subtilis, a strict aerobe which will not grow in the intestinal tract. In this situation where an organism growing in the environment fails to colonize the animal there may well be no effect of contamination on the parameter being studied. For example, it is quite likely that certain nutritional experiments could proceed in the face of heavy environmental contamination. On the other hand, the ingestion of large numbers of organisms from the environment could affect immunological experiments markedly. This is a situation which immunologists have to live with to some extent even in the absence of detectable viable organisms in the isolator since even dead dietary organisms are antigenic (Wagner, 1959b). Indeed the maintenance of antigenically virgin animals may prove to be impossible (but see Chapter 10).

Situations D and E where colonization of the animal occurs without or with growth in the environment, constitute a more serious threat to the experiment. Nevertheless there are situations such as in work on growth depression of chicks where colonization of the gut by, for example, *Escherichia coli* (Forbes *et al.*, 1959) does not affect the growth rate of the birds. Of course in all of these situations where a positive sterility test is obtained the animals must cease to be regarded as germ-free but the point to be made is that although animals become contaminated they may, as far as the particular parameter being studied is concerned, behave as if they were germ-free. Bearing in mind the expense and time involved in producing these animals, it is sometimes worthwhile continuing experiments after the animals have become contaminated.

C. TRACING THE SOURCE OF CONTAMINATION

A thorough investigation of a breakdown in sterility can be a very time-consuming exercise. However, on the basis of the type of organism involved one can usually make some intelligent guesses and logical eliminations. For example, in our experience micrococci are usually the result of glove breaks resulting in the transfer of micrococci from the skin of the hands to the animal. Glove holes are a particularly effective method of isolator contamination since they occur at that point of the system which comes into contact with the animal. Other leaks remain remote from the animal and may occur without giving rise to detectable contamination—particularly if a positive pressure is maintained within the isolator (cf. p. 7). We had on one occasion a nylon isolator which, although leaking, remained free from contamination for 24 days.

However, micrococcus contamination can arise in other ways. Early attempts to procure germ-free piglets often failed due to contamination with micrococci (Meyer *et al.*, 1963; R. Kenworthy, personal communication). The contamination occurred during the surgical delivery of the piglets and arose from micrococci which resided in the sebaceous glands of the sow's skin and were thus protected from the disinfectant used to sterilize the surface of the skin. The problem was overcome by replacing the scalpel used for the initial incision by a cautery.

Other useful guide lines which may be employed in tracing bacterial contamination of isolators are as follows. A. Spore-forming bacteria are relatively heat resistant compared with non-sporing species and are so ubiquitous that it is difficult to imagine an autoclave failure which did not lead to their being found in the isolator. However, because of their widespread distribution, when contamination with sporeformers does occur it is very difficult to suggest where they might have come from, although failure of heat sterilization should be suspected first. B. Organisms which do not grow at 37° C are unlikely to have come from the dam. C. In the case of chickens, contaminations with faecal type organisms such as *Streptococcus faecalis* or *E. coli* are most likely to have originated from inadequate sterilization of the eggs.

II. Methods Used for Routine Control

Although it is sometimes possible to detect contamination of an isolator by smelling the exhaust air or just by visual examination these are not satisfactory methods for the routine testing of germ-free isolators and a more extensive range of tests must be adopted. The basic requirement for cultivation of all potential contaminants is a good recovery medium. This can be an artificial medium, as in the case of bacteria, or living cells, as in the case of obligate parasites such as viruses. The tissues of the germ-free animal and its excreta are the best recovery media and smears of these materials showing large numbers of organisms usually indicate contamination. However, one should be on one's guard against diets containing a large number of dead organisms which can pass through the intestinal tract intact and give the impression that the animal is contaminated.

The methods used for establishing germ-freeness of an isolator range from visual examination by naked eye (for arthropods) through microscopic examination of wet mounts and stained preparations (for worms, protozoa, bacteria and fungi) to cultural tests on artificial media (for bacteria and fungi) and in living cells such as chick embryos, animals and tissue culture (for the detection of rickettsiae and viruses). The last mentioned may involve several passages before positive cultural tests are obtained.

Although this complete range of tests is necessary to establish the absence of all known forms of life, in most cases it is quite impracticable. In reality, because of the specialized facilities needed for the cultivation of viruses and rickettsiae these are not always tested for. It is desirable therefore to use animals derived from healthy stock to minimize the risk of virus contamination. Because of this limitation imposed by lack of facilities all that it is possible to say about the viral status of most animals is that they are free from symptomatic viruses.

Arthropods, parasitic worms and protozoa are more likely to be carried over from the parent animal than to be introduced subsequently. Wagner (1959a) failed to find any evidence of protozoal contamination in germ-free rats, mice, guinea pigs, rabbits and chickens despite the fact that protozoa were present in the dams of these animals. Nor was he able to find arthropods in any of the animals he examined. Nematodes appear to be a particularly common contaminant of germ-free dogs (Phillips, 1960; Griesemer and Gibson, 1963), the infection being acquired *in utero*. However, worm infestations were not found in the other species of germ-free animals examined by Phillips (1960). Thus the most likely cause of contamination at any stage subsequent to delivery would seem to be bacteria and these can usually be employed as an index of breakdown. They are extremely resistant organisms in many ways—not only to heat but to irradiation and chemical disinfection—so that if breakdown of one of these processes occurs it is most likely that bacteria will be amongst the organisms introduced. In addition they are commonly found in air and will also be introduced through leaks in the apparatus.

A. CULTURAL METHODS

The routine employed for the detection of bacteria and fungi in our isolators is based on that suggested by Wagner (1959a). Material from the intestine, excreta and drinking water is collected on swabs and treated as follows: (i) Inoculated within the isolator into tubes of thioglycollate medium and a general purpose medium—we use tryptic soy broth (Difco Laboratories, Detroit, Michigan, U.S.A.)—which are subsequently removed and incubated at 37° C aerobically. (ii) Removed from the isolator and inoculated on to the surface of plates of (a) tryptic soy agar with and without horse blood which are incubated at 37° C aerobically and anaerobically, and at 25° C aerobically; (b) potato dextrose agar which is incubated at 25° C aerobically. Many workers also include tests incubated at high temperatures (45–55° C) but the value of this is doubtful since in order for an organism to be present in detectable numbers in the isolator it must grow at 37° C or below.

In our experience, which is largely confined to tests on chickens, these procedures have always been adequate as judged by the fact that we have never had an occasion when stained smears showed the presence of large numbers of organisms which were not subsequently isolated. Contaminants that have been detected include moulds, yeasts, streptococci, micrococci, clostridia, coliforms, pseudomonads and aerobic sporeformers. A similar test used at Lobund for routine examination of rat, mouse and chicken colonies proved to be successful in detecting 99% of all the contaminations encountered (Anon, 1964).

The inclusion is advocated of media containing blood because blood is a useful and readily available growth stimulant for some bacteria. However, because its sterility cannot be guaranteed and because of its possible inhibitory properties towards some bacteria, it may give anomalous results. It is advantageous to use fluid thioglycollate medium since this can be inoculated *in situ* in the tank and reducing conditions, which facilitate the isolation of strict anaerobes, can be established quickly. By contrast there is some lag in the inoculation of swabs on to plates, but growth on solid medium often gives a better indication of mixed contamination than does broth. Plates of medium should be freshly poured to prevent the growth of micro-colonies which will subsequently be spread when the swab is cultured.

The material examined should be concentrated on the animal itself and its immediate environment. There seems little point in swabbing surfaces remote from the animal since any contamination in these sites is likely to be small and not of any consequence. All that is likely to be achieved is the transfer of a dormant contaminant into a situation where it can grow and cause overt contamination. With this in mind samples should be taken from the alimentary tract and skin of the animal and the water, diet and excreta. Whole animal culture at the end of the experiment may also be of use although Reyniers *et al.* (1949) stated that they had never found a contaminant in the tissues which was not also detectable in the gut.

B. MICROSCOPIC EXAMINATION

In addition to cultural tests, smears are made and stained by Gram's method. This practice is a safeguard against a situation where contamination is caused by an organism which will not grow on the media used. In the event of organisms being seen in the smear but not growing in or on the media, tests must be performed to decide whether these are merely dead organisms passing through the alimentary tract or whether they are organisms which require special media for *in vitro* cultivation. Usually a real contaminant which is colonizing the gut is seen in large numbers in the smear and the numbers increase from the anterior to the posterior ends of the intestinal tract. This type of increase can occur to a limited extent with dead organisms due to the concentration of the intestinal contents following absorption of soluble material (Reyniers *et al.*, 1949). This type of apparent increase would be less marked than that which occurs when an organism really becomes established in the gut. The other possibility which could arise is that of a fastidious organism growing only in the environment and passing through the intestinal tract without multiplying. Such a situation would be difficult to distinguish from the situation where dead organisms are ingested unless positive cultural tests are obtained by the use of special media. In all cases isolation of the organism should be attempted on special media. The type of organism seen in the smears will suggest the type of media which would be most suitable. For example, Gram positive bacilli in the smear would suggest the use of media suitable for clostridia or lactobacilli, whereas the presence of Gram negative bacilli might suggest the use of media known to be suitable for bacteroides. The use of media supplemented with germ-free gut contents may prove useful in isolating this type of organism.

C. PREVENTING ADVENTITIOUS CONTAMINATION

Sterility testing procedures should be attended by a scrupulous aseptic technique and a meticulous attention to detail. It should be remembered that careless sterility testing can not only give a false impression of the germ-free status of the isolator but can be a means of introducing contamination into it. Procedures for introducing sterility test material into the isolator should be checked to ensure that the heat treatment given is sufficient to achieve sterility without being so severe as to destroy the nutrient properties of the medium. It is important that operations carried out outside the isolator (e.g. inoculation of swabs on to solid media) should be done under a hood, in the area of a bunsen flame or in a sterile room. We have adopted a procedure whereby the fluid media inoculated in the isolator are removed in sealed containers which are then incubated at 37° C without opening. This technique obviates the possibility of anomalous results due to extra-isolator contamination.

D. FREQUENCY OF TESTING

The frequency of testing is not of critical importance. Because germ-free animals are time-consuming and expensive to maintain, it is desirable to know as soon as possible when contamination has occurred so that the experiment may be discontinued. On the other hand sterility testing cannot be undertaken too frequently without the tail beginning to wag the dog. On balance once a week seems to be a reasonable frequency of testing.

E. MONITORING POLYCONTAMINATED ISOLATORS

Special problems arise when monitoring mono- or poly-contaminated animals. In these cases it is necessary to be able to detect a contaminant in the presence of other organisms. In such a situation the microscopic examination of stained smears takes on added importance. If culture is to be successful then it may be necessary to employ selective media to avoid inhibition of the contaminant on the recovery medium. If the contaminant is closely related to an organism which has already been established as a deliberate contaminant in the animals, the problem of detection may be almost insurmountable. Nor is it safe to assume that if the organism is so similar as to be almost indistinguishable it will not affect the result. One could envisage a situation where a pathogenic serotype of *E. coli* gained access to animals previously contaminated with a non-pathogenic strain. Such a situation would give a false impression of the effects of the strain under test.

III. Recording Results

The maintenance of sterility of isolators is the corner-stone of all gnotobiotic research. And yet all too often the type of sterility control adopted is not recorded when the results of the experimental work are published, making the significance of the results difficult to assess. It is imperative that published results of gnotobiotic work be accompanied by a statement as to the type of sterility control adopted thus giving an indication of the limitations of the technique used—or to put it somewhat paradoxically, the "degree of germ-freeness".

REFERENCES

Anon (1964). *Gnotobiotics Newsletter* **1** (2), 5–6.
Forbes, M., Park, J. T. and Lev, M. (1959). *Ann. N.Y. Acad. Sci.* **78**, 321–327.
Fuller, R. and Moore, J. H. (1967). *J. gen. Microbiol.* **46**, 23–41.
Gibbons, R. J., Socransky, S. S. and Kapsimalis, B. (1964). *J. Bact.* **88**, 1316–1323.
Griesemer, R. A. and Gibson, J. P. (1963). *Lab. Anim. Care* **13**, 643–649.
Luckey, T. D. (1963). "Germfree Life and Gnotobiology". Academic Press, London.
Meyer, R. C., Bohl, E. H., Henthorne, R. D., Tharp, V. L. and Baldwin, D. E. (1963). *Lab. Anim. Care* **13**, 655–663.
Phillips, B. P. (1960). *Lobund Rep.* **3**, 172–175.
Reyniers, J. A., Trexler, P. C., Ervin, R. F., Wagner, M., Luckey, T. D. and Gordon, H. A. (1949). *Lobund Rep.* **2**, 1–116.
Socransky, S. S., Loesche, W. J., Hubersak, C. and Macdonald, J. C. (1964). *J. Bact.* **88**, 200–209.
Wagner, M. (1959a). *Ann. N.Y. Acad. Sci.* **78**, 89–101.
Wagner, M. (1959b). *Ann. N.Y. Acad. Sci.* **78**, 261–271.

Chapter 3

Animal Production and Rearing

Part I Small Laboratory Mammals

JULIAN R. PLEASANTS

*Department of Microbiology, University of Notre Dame,
Notre Dame, Indiana, U.S.A.*

Other chapters describe the diets and equipment required to keep animals germ-free and healthy. This chapter will summarize the modifications in animal management which are required for the germ-free rearing of small laboratory mammals, specifically the rat, mouse, guinea pig and rabbit and for their maintenance during experiments. If there were no differences between germ-free and conventional animal husbandry, it would be sufficient to refer the reader to any of a number of excellent handbooks of laboratory animal husbandry (Lane-Petter, 1963; Short and Woodnott, 1963; Porter and Lane-Petter, 1962; Worden and Lane-Petter, 1957; Farris, 1950). There are definite differences in management, however, which are reflected in the fact that the first of these books includes a chapter on germ-free animals and some of the others include a description of germ-free techniques.

The most obvious difference in management between germ-free and conventional mammals is the use of surgical operations to obtain the original stock of any germ-free mammalian line. Surgically derived rats, mice and rabbits, but not guinea pigs, must also be fed by hand on artificial milk formulas or foster-suckled on an already germ-free mother, until they are old enough to eat solid diets. Since not everyone who rears or uses germ-free animals needs to start his own line, the techniques of surgical delivery and hand feeding will be described at the end of this chapter. Everyone who uses germ-free animals in experiments, however, must know how his animals were obtained and maintained and how their method of rearing affects their quality and potential usefulness. This will be discussed in Section I. Every user of germ-free small mammals also needs to know how the procedures used to maintain animals germ-free can affect, often without any awareness on our part, the quality of our animal husbandry and our animals. These possible effects will be discussed in Section II, so that present-day users of germ-free animals can avoid repeating the mistakes made by those of us who reared them originally.

I. Colony Production *vs* Surgical Derivation

Animals produced for sale or use may themselves be surgically derived or may be produced by natural reproduction within the germ-free system. If the latter is true, the colony may be a closed one or may receive frequent introduction of new surgically derived stock to maintain desired genetic control. Each of these systems of production carries its own advantages and disadvantages. Both the producer and the user must be aware of the way in which the system of production can affect the micro-biological, physiological and genetic status of the animal, as well as the economics of its production.

A. MICROBIOLOGICAL STATUS

Proving the absence of all microbial associates is an involved and expensive process. When surgically derived mammals are used for experiment, each litter so derived would require checking for the whole bacterial spectrum as well as for filtrable agents. Pollard (1965) showed that leukamogenic viruses can be transmitted from parent to offspring in "germ-free" mice and our ability to prove their presence by precipitating overt disease may require special procedures and prolonged observation. Phillips and Wolfe (1961) reported the consistent occurrence of a respiratory syndrome in surgically derived germ-free guinea pigs obtained from a particular conventional colony. The syndrome was not observed in guinea pigs surgically derived from a different conventional source. Thus, in principle, each new litter introduced into the germ-free system should be submitted to exhaustive tests to establish its germ-free status.

The danger works both ways. A surgically derived litter which is itself free of microbial associates could acquire an associate from a "germ-free" foster mother which carries a vertically transmitted virus. Therefore, in theory, surgically derived litters of new strains should be hand fed, but the difficulty of hand feeding mice has led to the practice of foster-suckling new strains of mice on established germ-free lines, all of which are now known to harbour a leukamogenic virus. Since it appears that the new strains also possessed the leukamogenic virus before birth, they did not acquire any new associate through foster suckling. Nevertheless, hand-feeding would still be the ideal procedure for initiating new lines of germ-free mammals.

Because of the time and money involved in viral tests and exhaustive bacterial tests, they may be much more readily applied to a colony operation, in which new animals are obtained by reproduction within the system. By the criterion of microbiological certitude, the ideal germ-free animals come from a closed reproducing colony, or from a colony which is thoroughly tested after each introduction of more surgically derived stock, as well as routinely tested for microbes which might enter through a break in the system. In such a method of production there is time to build up an extensive microbiological "pedigree" for the animals which are used in experiments.

B. PHYSIOLOGICAL STATUS

Surgically derived germ-free rats, mice and rabbits must be hand-fed on sterilized milk substitutes, unless they can be foster-suckled on an existing germ-free line. So far, hand-fed germ-free rats and mice have shown slower growth and fur development during the suckling period than mother-suckled animals and have usually shown greater caecal distention (Pleasants, 1959; Gordon, 1959). For these species, therefore, the most nearly normal weanling animal is the one naturally born and normally suckled within the germ-free system. Hand-fed germ-free rabbits have shown more nearly normal growth rates and fur development (Reddy et al., 1965) than similarly reared rats and mice. The serious intestinal and renal abnormalities observed in the first germ-free rabbits (Wostmann and Pleasants, 1959; Luckey, 1963) were apparently due largely to the solid diets then used and have been alleviated by recent changes (Reddy et al., 1965). Therefore, hand-fed germ-free rabbits, by gross observation, are physiologically similar to those naturally born and normally suckled, but from the immunological viewpoint, they suffer from the disadvantage of exposure to heterologous (bovine) milk proteins (Wostmann, 1961).

The guinea pig occupies a paradoxical position. Surgically derived germ-free guinea pigs which received no milk substitute maintained normal caecal size for nearly 6 months when maintained on diet L-477 autoclaved with 25% added moisture (see Chapter 4), although distention gradually increased in later life. Naturally born, normally suckled germ-free guinea pigs had, however, developed caeca about 5 times normal size by 5 months of age (Pleasants *et al.*, 1967). From the point of view of anatomical normalcy, germ-free guinea pigs develop better without maternal milk. If they were to be routinely obtained by colony production, it might be advisable to separate them from their mothers within the first few days after birth. As far as we now know, surgical derivation and subsequent deprivation of maternal milk seem to have no physiological disadvantage for the germ-free guinea pig.

C. GENETIC STATUS

As mentioned earlier (Chapter 1), proper genetic management has been a serious problem in germ-free animal colonies, either because the colonies were not large enough for proper genetic management, or because it was feared that the frequent transfers of animals required for genetic management would spread contamination through a colony before it could be detected. This is what happened when a bacterial species difficult to culture from faeces spread throughout an entire colony of Lobund rats as a result of animal transfers carried out before the presence of the microorganism had been proved (Wagner, 1959). Genetic management has therefore represented a compromise between the standards set by the geneticists and the desire to avoid contamination risks. In some cases, luck favoured the compromise, as in the case of the Lobund Swiss-Webster mouse colony, which started from a single pair taken surgically from a random-bred colony and has bred into the 28th generation. However, in the case of germ-free guinea pig and rabbit colonies, also restricted in original numbers, there was a decline in fertility in later generations and some guinea pigs showed syndromes which could be traced to a double dose of genes from one male ancestor.

It is obvious that the full quality and value of the germ-free animal will not be attained under conditions of colony production until genetic standards are rigorously followed, in terms of original colony size and subsequent mating patterns, so that the animals will be comparable from isolator to isolator and from year to year. If techniques for keeping the animals germ-free are not dependable enough for such management, even though adequate for producing large numbers of animals, then improvement of techniques must be the first order of business. The pattern of genetic management would be the same for germ-free as for conventional animals (Lane-Petter, 1963) but by careful planning and genetic advice this goal could be attained with as few risks as possible of spreading contamination along with the genes.

D. ECONOMICS OF PRODUCTION

The demands of hand-feeding for surgically derived germ-free rats and mice make colony production economically as well as physiologically superior. Germ-free guinea pigs, however, require no hand-feeding and apparently are of better quality if they do not have maternal milk. Furthermore, the long gestation period and small litter size of guinea pigs necessitate a larger breeding stock to attain a given level of production than would be true of rats and mice. In our experience, second generation germ-free guinea pigs produced 0·7 young per month per female of breeding age (Pleasants et al., 1963a; Pleasants et al., 1967). A colony large enough for adequate genetic control and extensive production would make heavy demands on isolator space and therefore be relatively expensive. Thus, the primary reason for maintaining a reproducing germ-free guinea pig colony would not be physiological or economic but microbiological. An experimenter using surgically derived guinea pigs would have to decide whether he could submit them to adequate testing for his own purposes before actual use. While such animals may be useful for most purposes, some laboratories should maintain reproducing guinea pig colonies as a test of dietary adequacy and a test for the possibility of vertical transmission of viruses.

Germ-free rabbits occupy an intermediate position While surgically derived young require 3 weeks of hand feeding, the labour requirement is less than that for rats and mice because the animals take the nipple readily after the first day and because feeding need not be so frequent. Success in hand-rearing seems to depend on practice and has attained the level of weaning 75% of the young started (Pleasants et al., 1964). Colony production is expensive in terms of isolator space because adult breeders require separate cages and because the number of young produced per female is less than that for rats and mice, despite proverbs to the contrary. The Dutch strain used at Lobund reached about 2 kg adult weight, attained breeding age at 4 to 5 months and had litters averaging about 4 in number. Heavier breeds would weigh twice as much, take twice as long to reach breeding age and probably produce twice as many young per litter. Some experiments would call for the larger breeds.

Otherwise, small rabbits may be better accommodated in the usual types of isolators now available. We plan to start an inbred line of Dutch-size rabbits in order to minimize some of the genetic difficulties inherent in a colony of limited numbers. So many factors are involved in germ-free rabbit rearing that no clear-cut economic advantage can be seen for either colony rearing or surgical derivation. In that case, the physiological and microbial advantages of colony production should probably be decisive.

II. Meeting the Requirements of Good Animal Husbandry under Germ-free Conditions

A. ATMOSPHERIC CONDITIONS

This section is necessary because experimenters may not realize that the procedures adopted to maintain germ-free conditions can alter the physical and chemical environment of the animals without appearing to do so. The microclimate inside the cages where the animals actually live can be definitely affected by the procedure of wrapping a plastic or metal isolator around a group of cages. Movement of air across the cages changes from the mass movements found in animal rooms to a channeling system. Baker (1963) has shown that cage temperatures and relative humidity inside an isolator tend to run higher than those in animal room cages when the room temperature is the same. In metal isolators, the lights used may cause an unexpected increase in cage temperatures. Individual cages, according to their position in the isolator, might show more extreme deviations. Such changes, although small, could affect metabolic rates and possibly fertility.

Besides this chronic effect, temporary but sometimes drastic changes in the atmosphere result from the introduction of supplies into a germ-free isolator. When animals are raised in metal isolators with attached supply autoclaves, the temperature of the air inside the animal cages may rise above body temperature during autoclaving of supplies and may remain above optimum levels for hours afterwards (Miyakawa, 1967). As Miyakawa has shown, the effect can be much reduced by proper design. Since the introduction of plastic isolators, the most drastic change in the microclimate of the animals results from the introduction of peracetic acid fumes along with supplies. Heneghan and Gates (1966) have shown that such exposure constitutes an acute stress for germ-free dogs. It is not yet known how seriously other species are affected or what the long range effects are. Although peracetic acid belongs to a class of compounds which includes carcinogens (Van Duuren et al., 1963) it has not itself been implicated in carcinogenesis. Ethylene oxide as a sterilant for bedding for germ-free C3H mice has been linked retrospectively to an increase in tumor incidence (Reyniers *et al.*, 1964).

There is sufficient evidence for concern about effects of chemical sterilants on germ-free animals to warrant considerable effort to avoid such effects. Isolators chemically sterilized should be thoroughly ventilated before animals are introduced into them. The passage of peracetic acid fumes into an isolator from a sterile cylinder transfer is somewhat minimized by the fact that air flows out of the isolator through the cylinder filter during transfer. The fumes can be completely eliminated by ventilating the connecting sleeve overnight before opening the isolator, as described by Heneghan and Gates (1966). The greatest exposure to fumes comes from the use of a central supply isolator for many other isolators, since there is usually no time for ventilating the connecting sleeve and no air outlet in the sleeve to direct air from the isolator out through the sleeve.

Another possible hazard to life in a germ-free environment was discovered during research on germ-free plants (Fujiwara *et al.*, 1967). Plants became chlorotic in vinyl isolators when certain kinds of plasticizers had been used in manufacturing the vinyl film. Plasticizers used in making sterilizing membrane filters have also been found to affect tissue culture growth (Cahn, 1967). There has not yet been time to investigate the possibility that plasticizer vapours affect animal life.

All of the above considerations show the importance of designing isolators to assure adequate fresh air of the proper temperature and humidity in the immediate vicinity of the animals and to prevent contamination of this air with harmful chemical agents.

B. OVERCROWDING

Minimum standards for housing space have been set for conventional animals, but workers with germ-free animals may be tempted to believe that these standards are dictated more by microbiological than by psychological considerations. They may therefore be tempted to crowd germ-free animals, particularly in view of the great expense of germ-free space. The effects of overcrowding in the first germ-free rats were revealed in larger adrenal glands, smaller thymus glands and lower growth rates as compared to later animals (Gordon and Wostmann, 1960). Some experimenters and producers of germ-free animals find it psychologically impossible to remove germ-free animals which cannot be immediately used, but are crowding all the animals in the isolator. To maintain animal quality, one must be as ruthless with excess germ-free animals as with excess conventional animals. Although it has never been proved by a controlled experiment, some of us have the impression that germ-free animals are more sensitive to over-crowding than conventional animals and are more affected by aggressive activity if the sexes are not separated early enough. In research in general, there is a tendency to ignore the accumulating evidence about the effects of housing male rats or male mice in groups (Thiessen, 1964). In germ-free research, because of the scarcity of germ-free space, the temptation is even greater.

C. UNEXPECTED MANAGEMENT PROBLEMS

The sterilization of food pellets may create such a hard surface that pre-weaning rats and mice will be unable to eat them. The result will be slower growth of young during the last week of nursing and heavier demands on the nursing mother. The pellets may need to be broken open or moistened for nursing litters 14 days old and older. An unexpected hazard to life in young germ-free guinea pigs is their unusually strong tendency to suck the anus of other guinea pigs. This can lead to anal prolapse, followed by blockage and rupture of the colon. When this tendency is pronounced, the young may have to be separated from each other for several weeks.

III. Surgical Delivery

Germ-free animals may be delivered by hysterotomy (Caesarian section), in which the uterus is incised *in situ* and the young taken out individually, or by hysterectomy, in which the entire uterus is removed into the isolator before the foetuses are dissected from it. The details have been described in a number of publications. The methods for rats and mice have been given by Reyniers *et al.* (1946) and Gustafsson (1948). Various methods for guinea pigs have been recommended (Miyakawa, 1955; Miyakawa *et al.*, 1958; Tanami, 1959; Phillips *et al.*, 1959 and Abrams *et al.*, 1960). Rabbits were delivered at Lobund by the procedure used for guinea pigs, using local anaesthesia along the midline. The methods need not be elaborated here, since each experimenter will decide which is most convenient for his own use. In the case of these small animals, any of the published methods seems to produce bacteria-free young in the great majority of the operations.

Some general principles need to be emphasized. The timing of the delivery is very important for hand-fed animals, since slightly premature young seem to have much more difficulty in digesting artificial milk formulas than those delivered at term. For rats and mice the method developed by Reyniers and Lorenc (Reyniers, 1957) of waiting until one baby has been born naturally provides assurance of term deliveries. Rabbits deliver too rapidly to permit such a method of timing. They have been operated on at the 30th and 31st days of gestation and young have been successfully reared in both cases but the more developed young were easier to rear. There is a recent report of successfully delaying parturition in mice by the use of hormone preparations (Camden *et al.*, 1966). These have also been used successfully for large species and may become the usual means of assuring term deliveries. Parturition in guinea pigs has been routinely timed by palpating the pubic spread according to the method of Glimstedt (1936) and Phillips (1959).

IV. Hand-feeding

The details of hand-feeding infant rats, and of stimulating them for the elimination of wastes, have been given by Reyniers *et al.* (1946), Gustafsson (1948), Pleasants (1959) and Pleasants *et al.* (1964). Mouse hand-feeding was described by Pleasants (1959). Earlier methods of rabbit rearing were described by Wostmann and Pleasants (1959) and by Luckey (1963). The most recent method was described by Pleasants *et al.* (1963b and 1964). Present rabbit diets, both milk formulas and solid diets, are described in Chapter 4. More than half of the rats and rabbits, but only a few per cent of the mice, could be successfully hand-fed to weaning.

Guinea pigs need no hand-feeding or milk formula. In fact, all available evidence indicates that germ-free guinea pigs have difficulty in digesting any milk, even the unaltered milk of their own germ-free mothers. Thus, milk formulas should be avoided but solid diets for germ-free guinea pigs need to be modified in water content during the early weeks of life to encourage intake. A variety of diets has been used for germ-free surgically derived guinea pigs (Miyakawa *et al.*, 1958; Tanami, 1959; Phillips *et al.*, 1959; Horton and Hickey, 1961; Newton and DeWitt, 1961). The diet used at Lobund, which has proved its adequacy for sustained reproduction, is L-477, described in Chapter 4. The value of a proper water content during sterilization has been shown by Zimmerman and Wostmann (1963) and by Pleasants *et al.* (1967).

The only detail which seems to have been omitted in previous descriptions of hand-feeding is the procedure for making the natural rubber nipples used for rats, mice and rabbits. At Lobund Laboratory these are prepared from Lotol 6982 U, Naugatuck Chemical Corp., Naugatuck, Conn., U.S.A., the same type of latex used to make nipples for human babies. It is obtained fresh every few months and kept refrigerated. Glass tubes of the proper diameter are drawn out in a flame to the desired shape

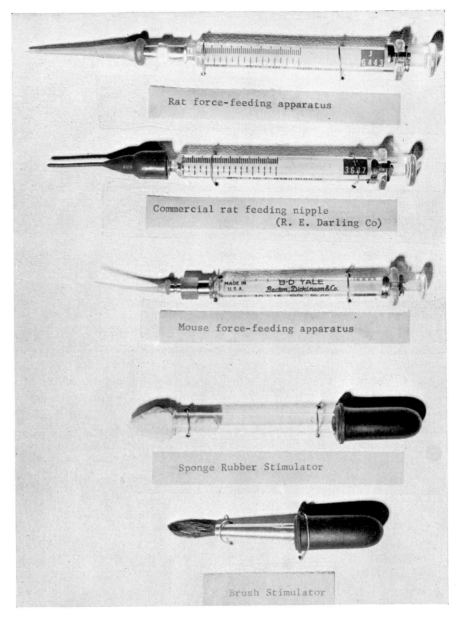

FIG. 1 Force-feeding nipples and stimulators used in the hand rearing of germ-free rats and mice.

(Fig. 1; see also Pleasants, 1959). They are held point upward in holes drilled in a board. Each form is dipped into the latex and then into 100% glacial acetic acid to precipitate the rubber. After this first coat is dry, the form is now dipped first into the acetic acid, then into the latex and back into the acetic acid. Experience will show how much time the form should be in the acid and in the latex in order to get a coat of proper thickness. A third coat may be applied if necessary. The form is allowed to dry, then soaked in distilled water for 1 hr to remove electrolytes and dried again. At this point, the lower portion of the latex coat can be rolled up to form a thick ring which helps to hold the nipple on the feeding tube or bottle. The forms are cured in an oven at 100° C for 1 hr, then boiled in water for 10 min to swell the rubber for easier removal from the form. The removed nipples are allowed to dry, then talcum powder is carefully applied both inside and out, to keep the rubber from sticking to itself. A hole of desired size is formed by puncturing or clipping the tip. The nipples are sterilized by autoclaving and used for feeding until they become swollen or soft.

The quantitative intake of diet has been a perennial problem in the hand-rearing of germ-free rats, mice and rabbits. The initial problem with rats and mice was to induce them to take enough diet on the *ad libitum* regimen of Reyniers *et al.* (1946). Therefore forced feeding was used by Gustafsson (1948), Pleasants (1959) and Miyakawa (1967). With rabbits, on the contrary, the problem was to keep them from eating too much too fast and therefore a feeding clamp was adopted to slow down intake (Fig. 2; see also Pleasants *et al.*, 1964). Nevertheless, with all three species, the quantity of food which could be safely taken was limited by the rate at which the diet was digested. This rate was usually so slow that there was only a narrow margin of safety between what could be safely fed and what the animal needed for growth. Excessive intake at one feeding led to regurgitation of diet (with the risk of diet aspiration) or to the sudden passage of undigested food from the stomach into the small intestine, where its further movement was even slower than usual.

The apparent difficulty of digesting protein in artificial formulas was associated with a precocious rise in stomach acidity in hand-fed rats (Wostmann, 1959), producing a stomach pH at which no proteolytic enzyme could work efficiently. The effort to by-pass protein digestion led to use of an amino acid-glucose-corn oil diet for infant rats (Pleasants and Wostmann, 1962; Pleasants *et al.*, 1964, see Chapter 4 for composition). This diet could be sterilized without chemical change by passage through a Millipore GS filter of 0·22 μ pore size. The two procedures used for introducing filtered diet into an isolator are shown in Figs 3 and 4, since these procedures have not been previously published. In the procedure of Fig. 3

the entire assembly was autoclaved and the diet was filtered under 20 lb/in.² nitrogen pressure. The neck of the flask was sealed in a flame and the sealed flask passed into the isolator through peracetic acid spray. In the procedure of Fig. 4, the pipe built into the entry port was sprayed with peracetic acid and the autoclaved filter was inserted into the upper end of the pipe. After 20 min, the inside stopper was removed and the diet was

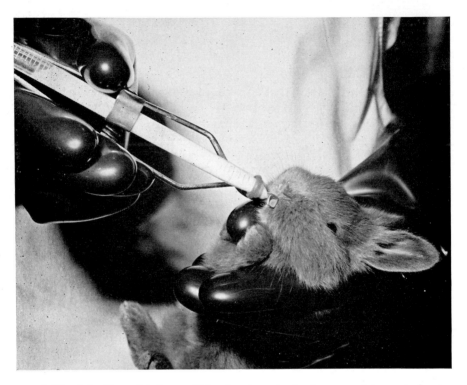

FIG. 2 Hand feeding of infant rabbit, showing nipple clamp used to control rate of intake.

filtered directly into a receiving vessel inside the isolator. A Tygon flare sealed to the top of the isolator has also been used in place of the rigid pipe shown in Fig. 4.

This water-soluble formula could be successfully used only after the 2 hr feeding interval previously used for milk formulas was replaced by a ½ hr feeding interval which minimized the osmotic effect of the new diet. This diet and technique have not yet been successful for infant germ-free mice but further improvements are now under investigation. It is hoped

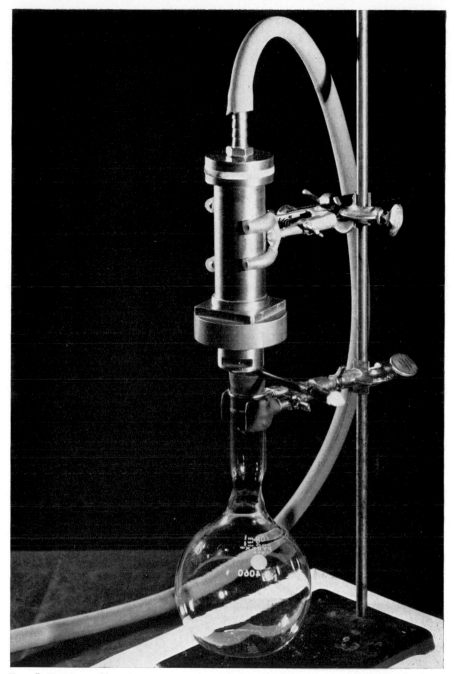

FIG. 3 Pressure filtration apparatus used for sterilization of water-soluble diet. Pipe extending through rubber stopper serves as pressure outlet. Constriction pre-formed in neck of flask facilitates heat sealing after filtration.

that this type of diet may eventually simplify the hand-rearing of other mammalian species and strains.

The philosophy of hand rearing which emerged from these experiences is this: while nature's methods of rearing infant animals—the composition of maternal milk, the natural mode of sucking and the normal feeding intervals—can be taken as points of departure for this kind of study, they should not be taken as the goals to be achieved in hand-rearing. The ultimate goal is a healthy weanling animal, rather than a particular

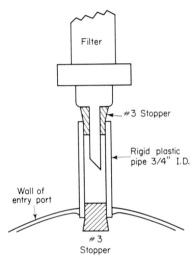

Fig. 4 Diagram of pressure filtration directly into germ-free isolator. After plastic pipe is sprayed with peracetic acid, the autoclaved filter assembly is fitted into pipe. After 20 min, inside stopper is removed to permit direct filtration into isolator.

procedure. The germ-free status of the suckling animal, the stress of hand feeding and the special properties of artificial milk formulas have already violated nature and made it impossible ever to duplicate natural conditions completely. Further violations may be needed to counteract the effects of those already made. Therefore it has been necessary to work out diet composition, feeding technique and feeding interval on a somewhat empirical basis, by analysis of their effects on the animals.

The germ-free suckling animal, for example, may be less capable than the conventional animal of hydrolysing lactose, despite the higher levels of lactase found in germ-free rats' intestinal mucosa by Reddy and Wostmann (1966). Bacteria could add considerably to the rate of hydrolysis as demonstrated for germ-free piglets by Schaffer et al. (1965). Absence of bacterial hydrolysis may be the reason for the earlier caecal distention seen in maternally-suckled germ-free guinea pigs (Pleasants *et al.*, 1967). Since whole cow's milk is already higher in lactose than the milk of rodent species, efforts should normally be made to reduce the proportion of lactose in total solids before feeding the formula to infant germ-free rodents. For specified pathogen-free infant rodents, such reduction might not be necessary, because they are usually given a lactose-fermenting flora at birth.

As another example, the feeding interval may have to be adjusted to the way that the milk formula curdles (or fails to curdle). The normal mode of curd formation has been described by Platt (1954). The liquid milk rapidly forms a solid clot by precipitating a new outer layer on the clot already present. This prevents both regurgitation of diet and rapid passage of the milk into the intestine. Since the usual formulas contain heterologous milk proteins (bovine), these may provoke a more severe response when reaching the lungs or lower intestine in unaltered form than would the same amount of homologous protein, as suggested by Coates and O'Donoghue (1967). Thus the feeding interval and amount may have to be adjusted by experience to prevent excessive build-up of uncurdled milk in the stomach. While it is entirely possible that infant rats, mice and rabbits do not need to be hand-fed during the night, the filling of their stomachs in preparation for this period should preferably be done gradually, either by frequent feeding (Pleasants, 1959) or by gradual filling of the stomach at a single feeding (Gustafsson 1948).

By following published procedures and gradually developing skill and judgment in hand-feeding, laboratories with the requisite staff should be able to hand-rear germ-free rats and rabbits consistently. It is hoped that present research into the hand-feeding of germ-free mice will produce techniques that can also give consistent success with this and other small species.

V. Summary

Early research on germ-free animal rearing necessarily concentrated on sterilization techniques and diet development, producing animals adequate for the more obvious comparisons of germ-free and conventional status. Recent developments, however, have shown that special consideration must be given to the demands of animal husbandry under germ-free conditions in order to maintain animal quality and permit more subtle discriminations of differences between germ-free and conventional animals. Proper attention to genetic management, exhaustive microbial testing, microclimatic conditions and animal housing, can provide a germ-free animal that will give consistent results over a broad range of experimental uses.

REFERENCES

Abrams, G. D., Bishop, J. E., Appelman, H. D. and French, A. J. (1960). *Univ. Mich. med. Bull.* **26**, 165–175.

Baker, D. E. J. (1963). *Proc. Gnotobiote Workshop and Symposium,* Ohio State Univ., July, 1963, 68–80.

Cahn, R. D. (1967). *Science, N.Y.* **155**, 195–196.

Camden, R. W., Poole, C. M. and Flynn, R. J. (1966). 17th Meeting Animal Care Panel: Abst. 74.

Coates, M. E. and O'Donoghue, P. N. (1967). *Nature, Lond.* **213**, 307–308.

Farris, E. J. (ed.) (1950). "The Care and Breeding of Laboratory Animals." John Wiley and Sons, Inc., New York.

Fujiwara, A., Ohira, K., Chiba, K. and Konno, I. (1967). *Internat. Symposium on Germfree Life Research.* Nagoya, Japan, April, 1967, p. 47 (Abst.).

Glimstedt, G. (1936). *Acta. path. microbiol. scand.* Suppl. **30**, 1–295.

Gordon, H. A. (1959). *Ann. N.Y. Acad. Sci.* **78**, 208–220.

Gordon, H. A. and Wostmann, B. S. (1960). *Anat. Rec.* **137**, 65–70.

Gustafsson, B. (1948). *Acta path. microbiol. scand.* Suppl. **73**, 1–130.

Heneghan, J. B. and Gates, D. F. (1966). *Lab. Anim. Care* **16**, 96–104.

Horton, R. E. and Hickey, J. L. S. (1961). *Proc. Anim. Care Panel* **11**, 93–106.

Lane-Petter, W. (ed.) (1963). "Animals for Research", Academic Press, London.

Luckey, T. D. (1963). "Germfree Life and Gnotobiology", Academic Press, New York.

Miyakawa, M. (1955). *Japan J. med. Prog.* **42**, 553.

Miyakawa, M. (1967). U.S.-Japan Seminar on Gnotobiotic Technology. Nagoya, Japan, April, 1967 (Abst.).

Miyakawa, M., Iijima, S., Kishimoto, H., Kobayashi, R., Tojima, M., Isamura, N., Asano, M. and Hong, S. C. (1958). *Acta path. jap.* **8**, 55–78.

Newton, W. L. and DeWitt, W. B. (1961). *J. Nutr.* **75**, 145–151.

Phillips, B. P., Wolfe, P. A. and Gordon, H. A. (1959). *Ann. N.Y. Acad. Sci.* **78**, 116–126.

Phillips, B. P. and Wolfe, P. A. (1961). *J. infect. Dis.* **108**, 12–18.

Platt, B. S. (1954). *Proc. Nutr. Soc.* **13**, xvi.

Pleasants, J. R. (1959). *Ann. N.Y. Acad. Sci.* **78**, 116–126.

Pleasants, J. R. and Wostmann, B. S. (1962). *Proc. Indiana Acad. Sci.* **72**, 87–92.

Pleasants, J. R., Zimmerman, D. R., Reddy, B. S. and Wostmann, B. S. (1963a). *Proc. Gnotobiote Workshop and Symposium,* pp. 36–50B, Ohio State Univ., July, 1963.

Pleasants, J. R., Zimmerman, D. R. and Wostmann, B. S. (1963b). *Lab. Anim. Care* **13**, 582–587.

Pleasants, J. R., Wostmann, B. S. and Zimmerman, D. R. (1964). *Lab. Anim. Care* **14**, 37–47.

Pleasants, J. R., Reddy, B. S., Zimmerman, D. R., Bruckner-Kardoss, E. and Wostmann, B. S. (1967). *Z. Versuchstierk.* **9**, 195–204.

Pollard, M. (1965). "Viral Status of 'Germfree' Mice." *Natn. Cancer Inst. Monogr.* No. **20**, 167–172.

Porter, G. and Lane-Petter, W. (1962). "Notes for Breeders of Common Laboratory Animals", Academic Press, London.

Reddy, B. S. and Wostmann, B. S. (1966). *Archs Biochem. Biophys.* **113**, 609–616.

Reddy, B. S., Pleasants, J. R., Zimmerman, D. R. and Wostmann, B. S. (1965). *J. Nutr.* **87**, 189–196.

Reyniers, J. A. (1957). *Proc. Anim. Care Panel* **7**, 9–29.

Reyniers, J. A., Trexler, P. C. and Ervin, R. F. (1946). *Lobund Rep.* No. 1. University Press, Notre Dame, Indiana, U.S.A.

Reyniers, J. A., Sacksteder, M. R. and Ashburn, L. L. (1964). *J. natn. Cancer Inst.* **32**, 1045–1056.

Schaffer, J., Walcher, D., Love, W., Breidenbach, G., Trexler, P. and Ashmore, J. (1965). *Proc. Soc. exp. Biol. Med.* **118**, 566–570.

Short, D. J. and Woodnott, D. P. (1963). "The A.T.A. Manual of Laboratory Animal Practice and Techniques." Crosby Lockwood, London.

Tanami, J. (1959). *J. Chiba med. Soc.* **35**, 1–24.

Thiessen, D. D. (1964). *Tex. Rep. Biol. Med.* **22**, 266–314.

Van Duuren, B. L., Nelsen, N., Orris, L., Palmes, E. D. and Schmitt, F. L. (1963). *J. natn. Cancer Inst.* **31**, 41–55.

Wagner, M. (1959). *Ann. N.Y. Acad. Sci.* **78**, 89–101.

Worden, A. N. and Lane-Petter, W. (eds) (1957). The UFAW Handbook on the Care and Management of Laboratory Animals, 2nd edn. Universities Federation for Animal Welfare, London.

Wostmann, B. S. (1959). *Ann. N.Y. Acad. Sci.* **78**, 175–182.

Wostmann, B. S. (1961). *Ann. N.Y. Acad. Sci.* **94**, 272–283.

Wostmann, B. S. and Pleasants, J. R. (1959). *Proc. Anim. Care Panel* **9**, 47–54.

Zimmerman, D. R. and Wostmann, B. S. (1963). *J. Nutr.* **79**, 318–322.

Part II Large Mammals

C. K. WHITEHAIR

Department of Pathology, Michigan State University,
East Lansing, Michigan, U.S.A.

Plastic isolators and equipment have been devised during the past decade so that pigs and other large mammals can be procured and maintained free of demonstrable microorganisms. In basic research, interest in animals obtained in this manner has been stimulated by the need for a more uniform and reliable experimental animal maintained in a controlled environment and, from a more practical viewpoint, as a method of eliminating costly infectious diseases existing in these species. Adaptation of the procedures and techniques used to produce and maintain germ-free laboratory animals at the University of Notre Dame seem to be the logical procedure to rear these larger species. The technical details in rearing germ-free laboratory animals such as equipment, diets and methods of determining contamination have been explored in detail and summarized by Reyniers (1959) and Trexler (1961) (see also Chapter 1). The application of these procedures to collecting and maintaining germ-free pigs and other farm animals has been described by Smith (1961, 1966), Waxler (1961a) and Waxler *et al.* (1966) and only the broader implications of the subject will be discussed here.

I. Pigs

A. HISTORY

Among the larger species, the pig has received most attention as a subject for gnotobiotic studies. Early interest (C. K. Whitehair, R. Dawley and O. B. Ross, unpublished data, University of Wisconsin, 1943) was precipitated by a desire to eliminate an enzootic diarrhoeal infection in baby pigs fed on experimental rations. The baby pigs were collected from a sow by hysterotomy using sterile techniques and removed to an isolated, but not germ-free, environment. Six of eight pigs were raised to maturity using a simple ration based primarily on cow's milk. Further work was abandoned at this time due to the war. The procedure was reinvestigated in 1950 (Thompson *et al.*, 1952; Whitehair and Thompson, 1956) by collecting baby pigs by hysterotomy, rearing them in isolation and observing production performance during two reproduction-lactation cycles. No unusual difficulties were encountered and it was noted that pigs reared in this manner did not respond to the addition of antibiotics in their ration.

The hysterectomy technique (Young and Underdahl, 1951, 1953), adapted primarily for repopulating swine herds free from virus pig pneumonia and atrophic rhinitis, was simpler. The gravid uterus was removed from the sow just prior to parturition and passed through an antiseptic trap into a hood where the pigs were removed. The baby pigs were reared in individual isolation (Horsfall) units (Young and Underdahl, 1953). This technique was recommended and received much publicity as a procedure to eliminate infectious diseases in swine, especially atrophic rhinitis and virus pig pneumonia. The procedure seemed satisfactory and good results were obtained during the first few months of operation. Very soon, however, in many of the commercial laboratories there was an increase in infectious agents which was difficult to control. In addition, the quality of pigs was poor. Today, with one or two exceptions, these laboratories have discontinued operations. Many of the operators of these laboratories were of the opinion that the programme left the research laboratory too soon and that it was oversold as to simplicity of operation.

Numerous other simple techniques have been reviewed and discussed by Goodwin (1965) and Young (1964). These simpler techniques, including collecting newborn pigs into sterile bags, all have advantages and disadvantages and merit consideration depending on existing conditions and the objectives of the work. Results of our research have not given us much confidence in maintaining young animals free of pathogens using isolation techniques. The hysterectomy technique, involving as it does exposure of the uterus to air-borne pathogens, has also failed to reliably control microbial contamination in our experience.

1. TERMINOLOGY

One of the first terms that was applied to animals aseptically delivered and reared in a "clean" environment was "disease-free". It was soon evident that these animals were not strictly disease-free and this term gave way to the term "specific pathogen-free" (SPF), a term still in use today. This term also has shortcomings and, at times, may be misleading. It is generally understood that the term refers to animals free of specified pathogens, yet other pathogens may be present that are as harmful (or more so) to the health of animals and the results of research. In pigs the term was used originally to describe those believed to be free of atrophic rhinitis and virus pneumonia. However, pigs from some commercial SPF laboratories developed enteric infections that seriously impaired their health and value for research studies. The final results will dictate the terminology and choice of techniques. Currently a description of the technique of collecting, rearing and maintaining the animals should be given and no claim made as to freedom of pathogens until adequately proven. We prefer the term "gnotobiotic" to describe the animals used in our research programme. The gnotobiotic animal may be germ-free or it may have one or more known species of microorganisms in its environment. Emphasis at Michigan State University during the past 8 years has been on the "closed system" as described by Trexler (1961) for rearing germ-free laboratory animals and adapted for pigs by Waxler (1961b) and Waxler et al. (1966) and for lambs and goats by Smith (1961, 1966). This technique allows complete control of the environment in which the animals are maintained. It is entirely different and should not be confused or compared with SPF systems of isolating animals at birth.

B. TECHNIQUES AND EQUIPMENT

The initial stages, namely the delivery and rearing of germ-free piglets from a conventional sow, are essentially similar whether the animals are ultimately intended as gnotobiotes for use in research or as disease-free stock for repopulating a commercial herd.

Contamination of germ-free animals may come from two sources, the uterus (Dunne, 1966) or the environment. Therefore, dams that are free of infections that might be transmitted *in utero* should be used for the production of germ-free young. In addition to checking the herd and individual health history and conducting appropriate clinical testing for infection, some workers suggest administration of antibiotics to the dam

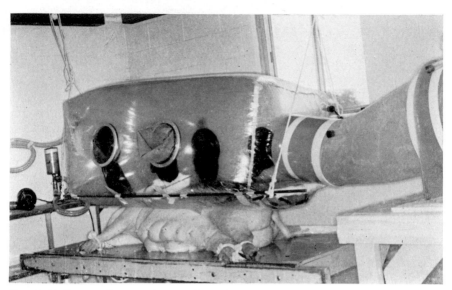

FIG. 1 Delivery of germ-free piglets. (A) Surgical isolator (left) and rearing isolator (right) are sterilized and equipped for the operation in advance. As the pigs are removed from the uterus they are passed into the rearing isolator on the right.

FIG. 2 Delivery of germ-free piglets. (B) The 30 cm fibreglass ring in the floor of the surgical isolator allows a more secure union between the skin of the sow and the isolator. It also facilitates the surgery in that forceps holding the incision open can be attached to the inside lip.

before surgery to reduce possibilities of contamination. Meticulous
planning is required ahead of time as the failure of any detail will ruin
the whole procedure. When an animal is placed in a closed system it is
sealed into that system with whatever contamination it possesses initially.

In our laboratory the plastic isolators, instruments, towels, cages and
other equipment are sterilized and made ready in advance (Fig. 1). The

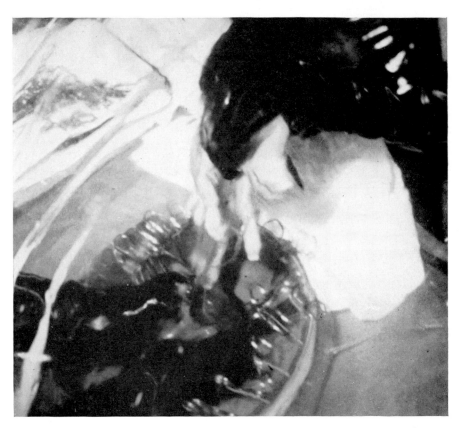

FIG. 3 Delivery of germ-free piglets. (c) An assistant removes the young, ligates
and severs the umbilical cord and passes the pig into the attached isolator.

isolators are cleansed and checked for leaks with a halogen detector after
introduction of a halogen-air mixture into the isolator. Cages, instruments
and diets are sterilized and introduced into the isolator where they are
sprayed with a 2% solution of peracetic acid.

Just prior to expected parturition, the left flank region of the sow is
prepared for surgery by shaving, cleansing with soap and water and

several applications of tincture of iodine. Intravenous administration of a tranquilizer followed by epidural administration at the level of the lumbosacral junction of 20 to 25 ml 2·5% procaine hydrochloride is our choice of anaesthesia. The sow is restrained on a surgical table in right lateral recumbency. Another application of tincture of iodine is made over the surgical area, the skin is wiped dry with sterile cotton and the

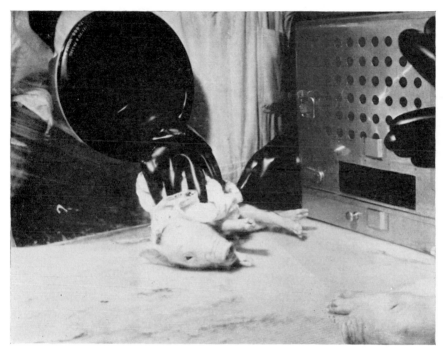

Fig. 4 Rearing isolator; the pigs are wiped dry, weighed, identified and placed in metabolism cages.

skin and the plastic film on the outside of a fibreglass ring (30 cm in diameter) is sprayed with a sterile surgical adhesive. The fibreglass ring in the floor of the isolator (Fig. 2) (Waxler *et al.*, 1966) allows a more secure union between the isolator and sow and reduces the incidence of contamination caused by the plastic pulling away from the skin. It also facilitates the surgical operation and the isolator is easily made ready for re-use by taping a new piece of plastic film over the ring. As the surgical isolator is lowered on to the flank of the sow, the surgeon, through the rubber gloves, adjusts the isolator to the location for hysterotomy. An electrical cautery unit is used to cut through the plastic film and skin of the sow. The cautery helps avoid contamination from hair follicles and skin glands. The incision is continued through the abdominal wall and uterus with a scalpel. As the surgeon opens the uterus the assistant on the opposite side of the isolator (Fig. 3) removes the young, ligates and severs the umbilical cord and passes the pigs into the attached rearing unit. As soon as the last pig is placed in the rearing unit, the lock is closed and the unit is removed to the rearing room. The incisions may be sutured and the sow allowed to recover or she may be destroyed. In the rearing units the pigs are wiped dry, weighed and placed in individual metabolism cages (Fig. 4).

C. DIETS

Diets for newborn germ-free pigs may be simple and yet satisfactory for maintenance as long as they remain free from contamination. Sterile homogenized cow's milk fortified with iron, copper and vitamin D is usually the diet of choice as it is convenient and easily prepared. It is sterilized by autoclaving in self-sealing 2-litre flasks. Baby pigs will consume milk in shallow pans starting 6 to 12 hr after birth. They are fed three times a day, with 120 to 240 ml of milk at each feeding. Additional vitamins and minerals may be supplied (Pleasants, 1959). Eaton (1963) reported improved results in pigs when glucose (8% of the caloric intake) and vitamin A (15,000 i.u.) per day were added to a basal ration composed of cow's milk. Pigs on this type of diet grow at about one-half the rate of pigs reared by the sow. The poorer growth is probably due to the lower energy content of cow's milk in comparison to sow's milk and the lower total milk consumption. Pigs given milk that simulated sow's milk grew faster (Whitehair and Waxler, 1963). The nutritive requirements of the newborn pig are fairly well established and a readily available sterile product supplying the nutritive requirements in their proper proportion and physical composition for the pig would be very useful. A commercial, sterile milk formula (Bordon's SPF-LAC) has been used with satisfactory results.

D. CONTROL OF REARING ENVIRONMENT

The newborn animal must be obtained free from contamination and then placed in a sterile environment in which it will be maintained. The problem of maintaining newborn animals germ-free for several weeks is even greater than collecting them from their mother. Attempts to raise germ-free or gnotobiotic animals in other than plastic isolators have not been dependable to control microbial contamination in our laboratory.

The hazards in rearing pigs in a "clean", but not germ-free, environment are even greater. The newborn, colostrum-deprived pig is very susceptible to a variety of infections. In addition the temperature of the environment, which is initially 35° C, the feeding habits of the young and milk type diets are all conducive to an increase or a "build-up" of pathogens which are difficult to control. This problem is not apparent the first time the equipment and facilities are used but increases with time. The failure to control these microorganisms is the primary reason that many commercial laboratories in the United States discontinued operations.

FIG. 5 Stainless steel metabolism cages for maintaining individual pigs to approximately 5 weeks of age. The cage should have no sharp corners and the door latch, feed tray and excreta tray must be adapted to handling with rubber gloves. The removable floor grid above the excreta tray is 1·27 cm mesh and is locked securely when in use.

The only method with which we have been able to maintain any degree of control of the microbial environment is by adaptation of the flexible film isolator used at Notre Dame to maintain continuous colonies of germ-free rats and mice (see Chapter 1). These isolators are not unduly expensive and can be made in about any size and to accommodate several pairs of gloves. The animals are maintained in individual stainless steel metabolism cages (Fig. 5) within the isolator. Stainless steel equipment is necessary because of the strong oxidizing properties of peracetic acid. The

isolator is 152 cm long × 61 cm wide × 61 cm high and will accommodate four metabolism cages 30·5 cm wide × 30·5 cm high × 46 cm deep
and leave sufficient room for feed supplies and handling equipment.
Larger isolators have been devised such as the 244 × 244 × 122 cm jacket
isolator (see Chapter 1) to accommodate a larger number of animals. In
experimental work on pigs a different treatment is usually given in each
isolator. Each will accommodate four pigs and we find no special advan-

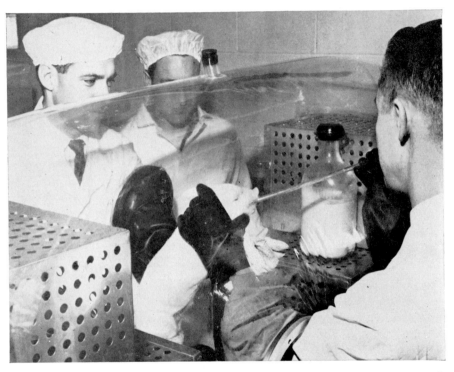

FIG. 6 Rearing isolator containing four metabolism cages, feed and experimental
equipment. Two sets of rubber gloves allow handling of the pigs and equipment for a variety of experimental work.

tages in the larger size isolators. Isolators containing animals infected
with a contagious disease, such as hog cholera (Weide *et al.*, 1962), may
be kept adjacent to isolators with germ-free animals without risk of
cross-contamination. Feed and other items are passed into the isolator and
refuse, tissue samples and empty containers passed out through a 45·7 cm
diameter lock sterilized with a 2% peracetic acid spray. The sterilized
metabolism cages may be introduced into the isolator through an entrance
port of this size. A pair of shoulder-length rubber gloves on each side of

the isolator allows caretakers to feed and handle the animals and manipulate the equipment (Fig. 6).

Specimens cultured for microbial contamination include rectal swabs and waste material from the cages. Determination of the microbial status of animals is usually conducted after they have been in the isolator for a week or 10 days. At this period there is generally a need to remove other material such as empty feed containers and blood samples from the isolator. The microbial status is determined by a variety of microbiological procedures as described by Wagner (1959). In our experience germ-free pigs readily adapted to conventional swine raising operations when transferred to a clean environment without exposure to conventionally reared pigs. However, a large mortality occurred (approximately 50%) when germ-free pigs were added to herds where evidence of infectious diseases were present.

E. MORPHOLOGICAL CHARACTERISTICS

A comparison of selected tissues from germ-free pigs has been made (Waxler, 1961b). Morphologically, the germ free pig is very similar to its conventionally-reared counterpart.

Histologically, the lymph nodes of germ-free pigs appear to have fewer germinal centres than nodes from farm-raised pigs. Also, the hepatic inter-lobular septa are less well developed in the germ-free pigs. At times the only indication of the border of the lobule was a tendency for the hepatic cells and reticular fibres to line up in these areas. The kidneys were heavier in germ-free pigs while the mandibular lymph nodes were lighter in germ-free than in farm-raised pigs.

Kenworthy and Allen (1966) noted marked differences in villus morphology between monocontaminated and multicontaminated pigs. In five-week old pigs monocontaminated with a strain of *E. coli* the villi of the small intestine were uniformly symmetrical and finger-shaped; in a duocontaminated littermate the uniformity was less pronounced and in litter-mates multicontaminated by housing in a conventional piggery the villi were considerably reduced in height, fused and club-shaped. The brush border was less clearly defined and the microvilli shorter in the multicontaminated compared with the monocontaminated pigs. These workers concluded that mucosal morphology is dependent, to some degree, upon biochemical interaction between bacteria and diet. Trapp *et al.* (1966) also noted marked differences in intestinal villus morphology between germ-free and pigs infected with the transmissible gastroenteritis virus (TGE). In germ-free pigs, the villi were long, slender and tapered to a rounded tip. The brush border was well-defined and uniform.

In the TGE infected pigs the villi were short, blunt, fused and the brush border was either shortened or not distinguishable. This information was believed to be of value in diagnostic pathology.

F. APPLICATIONS

1. IN RESEARCH

A wide variety of research projects may be conducted using the germ-free pig. These animals are of a size that permits collection of tissues and excreta for analytical work. Clinical observations can be made; they are usually available in the vicinity of most laboratories and they are susceptible to a variety of infectious agents. They are well enough developed at birth to consume readily a variety of experimental diets either to determine the nutritional requirements of the species under germ-free conditions or to conduct comparative nutritional studies. The germ-free technique seems to have special application in research on the pathogenesis of specific diseases and the role of microbial agents, either singly or in various combinations. Much interest has been manifested in using the germ-free animal to determine the contribution of the intestinal microflora to the welfare of the host animal, both in terms of nutrition and physical well-being. Studies on the interrelationship between nutrition and infectious diseases can be explored in detail with the germ-free animal. Many of the biomedical problems that may be investigated with the pig have recently been discussed (Bustad and McClellan, 1966).

2. IN DISEASE CONTROL

It is always difficult to measure accurately, in economic and other terms, a "negative" effect such as an infectious disease in livestock production. It is safe to assume that as large numbers of animals are concentrated in limited areas to meet efficient production goals, disease control measures will be the major production problem. The germ-free animal seems to be of value in three somewhat different areas: (1) as a subject in which to obtain more precise information on the pathogenesis of infectious diseases, (2) as a "monitor" to detect pathogens by exposing germ-free animals to animals or their tissues that may harbour subclinical infections and (3) as a technique by which diseases may be "by-passed" or eliminated by repopulating farms with "clean" animals.

While a voluminous amount of information is available on specific diseases, much more is needed in view of the magnitude of livestock disease problems. More knowledge is needed as to the physio-pathological role of the normal flora and its relationship with the known enteropathogens, the effect of dosage of infection, immunity and methods of disease spread.

The declining interest in the use of the germ-free pig in repopulating farms is due primarily to the lack of information on problems in operating herds started in this manner. While this technique has not met the expectations of some producers, it is believed to have accomplished much in the reduction of both the severity and incidence of atrophic rhinitis. The use of germ-free laboratory animals to repopulate conventional colonies of rats and mice has done much to reduce the problem of infectious diseases in these species (Hill, 1963). This was accomplished after much more information was available and many of the inherent problems were solved. In the laboratory animal the microbial status was established before repopulation of colonies was attempted. This has not been true in work on the larger species. The loss in production and income makes it expensive to repopulate swine herds and, unless superior results can be demonstrated, producers will not be inclined to change. Techniques must be devised that will assure producers of healthy breeding stock. One is the germ-free approach as described, which although perhaps expensive and time consuming does give better control over the possibility of contamination from the original animals.

II. Sheep, Goats and Cattle

Several species that in conventional circumstances are ruminants have been procured and maintained germ-free for a limited period of time. It is self-evident that, in the absence of a microbial flora, rumination cannot proceed. So far, these animals have not been maintained to the stage when rumination would normally be established.

A. TECHNIQUE

The first successful attempt to rear one of the ruminant species germ-free was that of Küster (1915) who procured young goats and reared one germ-free on sterile goat's milk for 36 days.

Smith (1961, 1966) has probably made the most extensive attempt to rear germ-free goats and lambs. Twenty-seven lambs and goats were procured and maintained in plastic film isolators. Fifteen were maintained germ-free for varying periods of time ranging from 13 to 127 days of age. Hysterotomy was the most successful and least difficult technique in procuring the young germ-free from the dam. Morter et al. (1964) reported the techniques for aseptically procuring calves and rearing them in isolation to establish a disease-free herd of cattle.

To maintain the larger species germ-free much of the foregoing discussion on pigs and laboratory animals is applicable. The type of isolator, sterilization techniques and methods of procuring the young are similar. The major problem, due to the size of the young, is the manipulation of the young from the uterus to within isolators under germ-free conditions. Sterilizing and getting the larger cages into the isolator is also difficult. The greater amounts of food required by the larger species presents problems in sterilization, entrance into the isolator and storage within the isolator. However, most of the larger mammals, since they are more developed at birth, will readily consume food that is offered.

B. MORPHOLOGY

On gross examination of a number of tissues from the germ-free goat or lamb, they appeared similar to those reared by conventional methods (Smith, 1966). There was no significant difference in the weight of various organs. Morphologically rumen papillae were present in the germ-free animals but the rumen was immature and lacked the vacuolization in the epithelium characteristic of the normal ruminant. The intestinal tract of the germ-free animal had an increased number of eosinophilic cells in the lamina propria and less connective tissue and cellular infiltration in the submucosa. There was evidence of less lymphatic development; splenic corpuscles were smaller and less frequent; less interlobular connective tissue in the liver and lower values for serum proteins, gamma globulin and haemoglobin in germ-free animals. The blood glucose values in the germ-free animal did not decline with age as in the normal ruminant.

C. APPLICATIONS

The establishment of germ-free larger species of livestock has both economic potential and basic research interest. The practical advantage is to establish herds and flocks free of the infectious diseases. The research interests lie in studying the pathogenesis of specific infections without the influence of extraneous factors. Also, the ruminant animal requires a symbiotic microflora to maintain an adequate nutritional status. The complex of bacteria and protozoa provides a source of water soluble vitamins and volatile fatty acids, the latter being a major source of the energy requirements. Cellulose digestion is mediated through the action of microorganisms in the rumen and microbial protein formed from non-protein nitrogen makes a significant contribution to the nutritional requirements of the host. In the germ-free "ruminant" the metabolism of single nutrients or a defined ration can be determined in the absence of microflora or in the presence of a specific flora. These studies will be essential to a full understanding of ruminant nutrition and physiology.

III. Summary

The germ-free pig and other large mammals have many uses in both basic and applied research in livestock production problems. Additional research is needed on techniques, equipment, diets and performance under practical conditions. The work is time consuming, expensive and, at times, frustrating but there may be no alternative in solving some of the problems of livestock production.

REFERENCES

Bustad, L. K. and McClellan, R. O. (eds) (1966). "Swine in Biomedical Research." Pacific Northwest Laboratory, Richlend, Washington.

Dunne, H. W. (1966). In "Swine in Biomedical Research" (L. K. Bustad and R. O. McClellan, eds), pp. 727–729, Pacific Northwest Laboratory, Richlend, Washington.

Eaton, B. G. (1963). Lab. Anim. Care Suppl. 13, 673–674.

Goodwin, R. F. W. (1965). Vet. Rec. 77, 1070–1076.

Hill, B. F. (1963). Lab. Anim. Care Suppl. 13, 622–623.

Kenworthy, R. and Allen, W. D. (1966). J. comp. Path. Ther. 76, 291–296.

Küster, E. (1915). Arb. K. GesundhAmt. 48, 1–79.

Morter, R. L., Herschler, R. C. and Brobst, D. (1964). Proc. Gnotobiotic Sym. and Workshop. Michigan State University East Lansing, Michigan, U.S.A.

Pleasants, J. R. (1959). Ann. N.Y. Acad. Sci. 78, 116–126.

Reyniers, J. A. (Chm. and ed.) (1959). "Germfree Vertebrates: Present Status." Ann. N.Y. Acad. Sci. 78, 1–400.

Smith, C. K. (1961). In "Application of Caesarean-Derived Animals to Disease Control in Livestock and Laboratory Animal Production", pp. 26–27. Michigan State University East Lansing, Michigan, U.S.A.

Smith, C. K. (1966). "The Derivation and the Characterization of the Germfree Ruminant." Ph.D. Thesis. University of Notre Dame, Indiana.

Thompson, C. M., Whitehair, C. K., MacVicar, R. W. and Hillier, J. C. (1952). Okla. Agric. Exp. Sta. M-P 27.

Trapp, A. L., Sanger, V. L. and Stalnaker, E. (1966). Am. J. vet. Res. 27, 1695–1702.

Trexler, P. C. (1961). Bio-Medical Purview 1, 47–58.

Wagner, M. (1959). Ann. N.Y. Acad. Sci. 78, 89–101.

Waxler, G. L. (1961a). In "Application of Caesarean-Derived Animals to Disease Control in Livestock and Laboratory Animal Production", pp. 13–17. Michigan State University, East Lansing, Michigan, U.S.A.

Waxler, G. L. (1961b). "Studies on Gnotobiotic Pigs." Ph.D. Thesis, Michigan State University, East Lansing, Michigan, U.S.A.

Waxler, G. L., Schmidt, D. A. and Whitehair, C. K. (1966). Am. J. vet. Res. 27, 300–307.

Weide, K. D., Waxler, G. L., Whitehair, C. K. and Morrill, C. C. (1962). J. Am. vet. med. Ass. 140, 1056–1061.

Whitehair, C. K. and Thompson, C. M. (1956). *J. Am. vet. med. Ass.* **128**, 94–98.

Whitehair, C. K. and Waxler, G. L. (1963). *Lab. Anim. Care* Suppl. **13**, 665–672.

Young, G. A. and Underdahl, N. R. (1951). *Archs Biochem. Biophys.* **32**, 449–450.

Young, G. A. and Underdahl, N. R. (1953). *Am. J. vet. Res.* **14**, 571–574.

Young, G. A. (1964). "SPF Swine." *Adv. vet. Sci.* **9** Academic Press, New York and London.

Part III Chickens and Quail

M. E. COATES

National Institute for Research in Dairying, (University of Reading)
Shinfield, Reading, England

I. Chickens

As a species suitable for production in a germ-free environment, the oviparous nature of the chicken gives it some advantage over mammals. Fertile eggs are cheap and plentiful; surgical delivery of the young is unnecessary and there are no problems of hand-rearing since the bird can fend for itself within a few hours of emerging from the egg. However, many investigators complain of a higher incidence of contamination in experiments with chickens than is usually experienced with mammals. Although the contaminant is not always detected at the beginning of an experiment, it is conceivable that organisms may be introduced with the eggs but remain dormant until conditions within the isolator become favourable for their multiplication.

A. PRODUCTION OF EGGS

1. INCIDENCE OF CONTAMINATION IN EGGS

The freshly-laid egg is generally regarded as a closed germ-free system for which the mechanical barrier of the shell and the bacteriolytic action of lysozyme in the egg white offer protection against bacterial invasion. Cottral (1952) expressed the opinion that the great majority of eggs are free from contamination, although it undoubtedly can occur, either endogenously by transmission of an infective agent from the dam

or exogenously by invasion through the shell. The transmission from dam
to offspring of certain viral and bacterial diseases of poultry (for instance,
leucosis and pullorum disease) is well recognized but congenital
organisms rarely occur in eggs from healthy birds.

In a study of the bacterial content of the reproductive tract in the fowl,
Harry (1963a) observed a very low incidence of bacteria in the ova and
oviduct of laying birds; thus it appears that the risk of inclusion of
organisms in the egg during the processes of albumen- and shell-
formation is low. In the vagina of naturally-mated hens he recorded an
appreciable count of organisms and this was markedly greater in number
and variety in birds that had been artificially-inseminated. Buxton and
Gordon (1947) have demonstrated with cultures of *Salmonella thompson*
painted on the shell surface of hatching eggs that warm, moist conditions
favour penetration of bacteria into the interior. The egg is therefore
most vulnerable to microbial invasion immediately after lay and the
organisms present in the vagina constitute a likely source of contaminants.
The authors observed little penetration during storage in a cool, dry
atmosphere but conditions prevailing during incubation were favourable
to bacterial invasion.

2. MANAGEMENT OF THE PRODUCTION FLOCK AND SELECTION OF EGGS

The question of egg and hatchery hygiene in relation to the spread of
egg-borne disease has been fully discussed by Harry and Gordon (1966).
Many of the principles they recommend might usefully be applied in the
production of eggs intended to hatch germ-free chickens. It is self-evident
that the parent flock must be free from transmissible disease. With proper
care, diseases of bacterial origin can usually be avoided. Although a few
"barrier-sustained" flocks, free from recognized viral diseases, are
maintained at specialized research stations, under ordinary conditions of
management freedom from endogenous viral contaminants is virtually
unattainable. Germ-free chicks hatched from commercially-produced eggs
are likely, therefore, to carry egg-borne viruses even though no clinical
signs of disease are apparent.

The major incidence of bacterial contamination is likely to arise by
microbial invasion through the shell, hence all precautions should be
taken to reduce the number of organisms to which the egg is exposed
during the period from laying to hatching. This implies the initial
production of clean eggs and their subsequent protection against unnecessary contact with microorganisms. In selecting eggs a close inspection
should be made and only visibly clean eggs, with good shells free from
cracks, should be chosen.

The housing and management of the parent flock can markedly

influence egg cleanliness; for instance, soilage by mud and excreta is more common in eggs from hens housed on deep litter or grass runs than on eggs from birds in wire-floored cages. Dust from litter also presents a hazard. Harry (1963b) found fifteen times as many bacteria on apparently clean eggs from hens on deep litter than on comparable eggs from birds in battery cages. Housing in batteries necessitates artificial insemination, with a consequent risk of increasing the microbial burden in the vagina, but the "verandah" system currently popular with commercial egg producers offers a satisfactory compromise. In this type of accommodation the flock is housed in a pen with a raised wire-screen floor, the birds' feet are relatively clean and the spread of dust and dirt is consequently reduced. A "rollaway" device, in which the eggs are laid on a sloping wire floor and can roll out of reach of the hen, is helpful. If conventional nest-boxes are used, the nesting material should be frequently changed; wood wool or paper shavings are likely to carry a lower microbial load than hay or straw.

Frequent collection of eggs reduces the time of exposure to soiling in the nest. They should never be scrubbed or treated with abrasives because such treatments might damage the protective cuticle. In commercial hatcheries, treatment of the eggs shortly after lay with disinfectant dips has proved effective in reducing contamination in hatching eggs. It is best done before the newly-laid eggs have cooled, and consists of immersion in a warm disinfectant solution, with gentle agitation, for a few minutes. Alternatively, the dipping procedure may be performed immediately before the eggs are set in the incubator. A number of preparations are marketed for the purpose and the concentration and time of treatment vary according to the manufacturers' instructions. Fumigation with formaldehyde vapour is an effective alternative. Before incubation, the eggs should be stored in a cool, dry, dust-free atmosphere for as short a time as is practicable.

Fumigation of the incubator with formaldehyde between hatches is a wise precaution. Fumigation during incubation has been recommended to control bacterial infections in hatcheries but this procedure must be applied with caution in view of reports of harmful effects at some stages in incubation. Formaldehyde is generated from a mixture of two parts potassium permanganate to three of formalin. Clarenburg and Romijn (1954) found no effect on the number or quality of the chicks hatched from eggs fumigated with 150 ml formalin per 100 ft^3 (53 ml per m^3) of incubator space at any time from pre-incubation to 72 hr incubation. Harry and Binstead (1961) showed that hatchability could be adversely affected if the eggs were exposed to this concentration on the third to the ninth days of incubation. At the Lobund Laboratories incubators are fumigated weekly using 35 ml formalin per 100 ft³ (B. A. Teah, private communication).

B. PRODUCTION OF GERM-FREE CHICKS

1. INTRODUCTION OF EGGS INTO THE ISOLATOR

The usual practice is to incubate the eggs in a standard incubator until about the eighteenth day of development when, after a further inspection for cleanliness, those containing live embryos are selected by candling and put through a decontamination process before being taken into the isolator. They are most conveniently treated in batches of six to a dozen, which can be packed into a net bag for ease of handling. The choice of disinfectant is limited, since the effective concentration and time of treatment must not be such as to allow penetration of the shell and consequent harm to the embryo. The customary procedure is a mild detergent wash to remove surface dust and to wet the shell, followed by immersion in a 1 or 2% solution of mercuric chloride for 8 min; a gentle massaging action ensures that the solution comes in contact with all parts of the shell. The solutions should be at the same temperature as the incubator (about 37° C); cooling at this stage would chill the embryo and, by contraction of air inside the shell, might cause surface organisms or germicide to be drawn in through the pores. The Reyniers design of isolator has the advantage that eggs can be passed directly from the mercuric chloride solution into the interior where the residue of germicide carried in on the shell can continue to act. In isolators with a built-in germicidal trap, much of the $HgCl_2$ is inevitably washed off when the eggs are passed in through the trap. At the Lobund Laboratories the eggs are candled on the twentieth day of incubation. The selected eggs are dipped in 2% peracetic acid, then immersed in an iodophor solution (0·08% available iodine) for 10 min at 37° C before they are passed into the isolator through a germicidal trap (B. A. Teah, private communication).

In plastic isolators with a sterile lock, eggs can be decontaminated in the lock by spraying with a 2% peracetic acid and detergent solution and leaving them exposed to the vapour for 30 min before passing them into the main compartment of the isolator.

2. HATCHING

Inside the isolator temperature and humidity must be carefully controlled during the remaining three days before hatching; wet- and dry-bulb thermometers should always be among the articles included in isolators intended for chicks. Temperature should be evenly maintained at 37–38° C. A few degrees increase is lethal to the developing embryo;

if it is allowed to fall much below 35° C a delayed hatch and, possibly, malformed chicks may result. Infra-red lamps provide a simple way of heating metal isolators but may scorch plastic film. Alternatively, isolators can be placed in a temperature-controlled room during the hatching period.

A greater range of relative humidity can be tolerated by the developing embryo, with the optimum about 65–70%. At low humidity the egg membrane becomes dry and hard and the bird has difficulty in pecking its way into the air space; those that do succeed often remain stuck to the shell. Very high humidity interferes with the establishment of pulmonary respiration and the chicks "drown" before they can hatch. Humidity can be raised inside the isolator by exposing pans of water or wet sponges. It is easily controlled in apparatus with the incinerator type of air sterilizer by dripping water on to the heated carborundum at a rate regulated to give the required level of moisture in the incoming air.

The most difficult problem in hatching eggs inside an isolator is likely to be the maintenance of adequate ventilation. The amount of air entering the isolator is limited by the size of the inlet and exit pipes and the resistance of the filters, hence its rate of flow may be insufficient to create turbulence and CO_2 may accumulate in the region of the eggs. The eggs should therefore be set as near to the fresh air intake as possible.

TABLE I

Effect of disinfection on hatchability of chicken eggs

Treatment	Place of hatching	No. of eggs	No. of chicks	Hatchability %
None	Incubator	4000	3000	75
2% HgCl₂ for 8 min	Incubator	1004	605	60
	Isolators	498	283	56

The disinfection process inevitably has an adverse effect on hatchability, partly because of displacement of the embryos during handling and possibly also by penetration of disinfectant through the shell. A further reduction may occur if environmental conditions within the isolator are not optimal. In the course of a series of experiments in this laboratory eggs were incubated in a standard commercial machine for eighteen days, then disinfected with $HgCl_2$ and either replaced in the incubator or passed into Gustafsson isolators. The percentage hatchability, together with that of untreated eggs incubated in the normal way, is recorded in Table I.

There was a marked reduction in the percentage hatched of eggs that had been disinfected and a' further small decrease inside the isolator. For reasons connected with other experimental work, the general level of hatchability was low on this occasion but with eggs of commercial origin, a hatch of about 70% can be expected inside the isolators.

C. REARING GERM-FREE CHICKS

As soon as the chicks dry after hatching they readily find their way to the sources of food and water and present few difficulties in their management. Temperature should be held at about 35° C for the first few days, then gradually reduced to 25° C or less over the next three

TABLE II

Weight at four weeks of germ-free and conventional chicks

Laboratory	Type of diet	No. of birds per group	Mean body weight (g) Germ-free	Conventional
Walter Reed Army Institute of Research	Soya/maize[a]	22	348	284
	Casein/starch[a]	22	358	290
	Casein/starch[a]	20	326	302
National Institute for Research in Dairying	Chick starter mash[b]	50	300	256
	Casein/starch	96	348	331

[a] Forbes and Park (1959).
[b] Coates et al. (1963).

weeks. As the birds grow, two main problems arise. First, the accumulation of dust and fluff from the diet and feathers may be sufficient to block the outlet filters; it is advisable therefore to fit a coarse filter, that can be frequently changed from inside the apparatus, over the outgoing air pipe. Second, humidity will almost certainly increase enormously through evaporation from droppings and water troughs and the chicks' habitual splashing of their drinking water. Although a high relative humidity does not seriously distress the birds, moisture condenses on the cooler parts of the isolator and may interfere with visibility. Chicks are not readily trained to drink from a drip device but the exposed surface in the drinking troughs can be reduced to quite a small area. Increasing the ventilation rate to carry away the increased load of water vapour is the most effective way of maintaining humidity at a reasonable level.

Germ-free chicks grow satisfactorily on commercial chick-starter rations or on laboratory diets of purified ingredients. Their nutrient requirements seem adequately met by the allowances recommended for growing chicks by, for instance, the National Research Council of the United States (1960) or the Agricultural Research Council of Great Britain (1963). There is evidence, cited in Table II, that the early rate of growth is faster in a germ-free than in a conventional environment. Young chickens make suitable subjects for studies of nutrition and metabolism. They are less convenient for long-term experiments because of their large body-size and slow rate of reproduction.

II. Quail

A. MANAGEMENT OF A LABORATORY FLOCK

The Japanese quail (*Coturnix coturnix japonica*) is a very small, rapidly maturing bird that may prove a useful alternative to the chicken in germ-free experiments. The adults weigh only about 120 g; the first egg is usually laid at about six weeks of age and full sexual maturity is reached at ten weeks. The production flock can be conveniently housed in the laboratory in a relatively small space. In their natural habitat they lay only in the spring but, given even temperature and ample artificial light, they can be kept in production for eleven months of the year. Further details of Japanese quail husbandry in the laboratory have been given by Woodard *et al.* (1965) and Miles (1966).

The precautions suggested above for the production of clean chicken eggs are even more important for quail eggs. The female does not nest, but usually deposits the egg in the run. As the shells are extremely fragile, the incidence of cracked and dirty eggs is high. It can be reduced by caging the birds on sloping floors, which allow the eggs to roll out of contact as soon as they are laid. Cracks and stains are difficult to detect because of the natural colour of the shell, which is covered with patches of brown pigment.

B. INCUBATION OF EGGS

The normal incubation time for quail eggs is seventeen days and they are usually taken into the isolator on the fourteenth or fifteenth day of incubation. Reyniers and Sacksteder (1960) recommend treatment with a detergent before incubation and disinfection for 5 min with 2% $HgCl_2$ solution immediately before passing into the isolator. In our experience, the detrimental effect of the disinfection procedure on hatchability is even greater with quail than with chicken eggs. Optimal incubation conditions for quail eggs have not yet been clearly defined but they seem similar to those for chicken eggs. A very high humidity is desirable while the birds are hatching and it is often helpful to spray the eggs lightly with water several times during the fifteenth and sixteenth day of development.

Quail have been bred in Reyniers isolators (Reyniers and Sacksteder, 1960); the end autoclave was adapted to serve as an incubator by surrounding it externally with a thermostatically-controlled heating coil.

III. Summary

The production and management of germ-free chickens and quail is relatively simple, although there is a serious risk of contaminants being carried into the isolator on the shells of hatching eggs. The risk can be greatly reduced by careful management of the production flock to minimize soiling of the eggs. Special hazards that arise in the rearing and maintenance of birds are the accumulation of dust and the development of a high relative humidity within the isolators. Germ-free chicks grow better than their conventional counterparts, but are not often used for long term experiments because they mature slowly and occupy a considerable amount of isolator space. The Japanese quail, because of its small size and rapid rate of reproduction, may become a useful alternative in experiments lasting beyond the first generation.

REFERENCES

Agricultural Research Council (1963). "The Nutrient Requirements of Farm Livestock." No. 1, "Poultry". Agricultural Research Council, London.
Buxton, A. and Gordon, R. F. (1947). *J. Hyg., Camb.* **45**, 265–281.
Clarenburg, A. and Romijn, C. (1954). *Wld's Poult. Congr. X, Edinburgh*, 214–217.
Coates, M. E., Fuller, R., Harrison, G. F., Lev, M. and Suffolk, S. F. (1963). *Br. J. Nutr.* **17**, 141–150.
Cottral, G. E. (1952). *Ann. N.Y. Acad. Sci.* **55**, 221–235.
Forbes, M. and Park, J. T. (1959). *J. Nutr.* **67**, 69–84.
Harry, E. G. (1963a). *Br. Poult. Sci.* **4**, 63–70.
Harry, E. G. (1963b). *Br. Poult. Sci.* **4**, 91–100.
Harry, E. G. and Binstead, J. A. (1961). *Br. vet. J.* **117**, 532–539.
Harry, E. G. and Gordon, R. F. (1966). *Veterinarian, Lond.* **4**, 5–15.
Miles, D. A. F. (1966). *J. Inst. Anim. Techns* **17**, 74–79.
National Research Council (1960). *Publs natn. Res. Coun., Wash.* No. 827.
Reyniers, J. A. and Sacksteder, M. R. (1960). *J. natn. Cancer Inst.* **24**, 1405–1421.
Woodard, A. E., Ablanalp, H. and Wilson, W. O. (1965). "Japanese Quail Husbandry in the Laboratory." Dept. of Poultry Husbandry, University of California, Davis, California, U.S.A.

Chapter 4

Nutritionally Adequate Diets for Germ-free Animals

BANDARU S. REDDY, BERNARD S. WOSTMANN AND
JULIAN R. PLEASANTS

*Lobund Laboratories, University of Notre Dame,
Notre Dame, Indiana, U.S.A.*

I. Introduction

Although the basic concept of germ-free animal research originated from Pasteur, extensive application was held back by lack of knowledge about the effects of diet sterilization on essential nutrients. It was only after World War II that the nutritional requirements of the most widely used experimental animals and the effects of diet sterilization were sufficiently understood to maintain large numbers of several species germ-free for a prolonged period of time.

The successful germ-free rearing of many animal species (Luckey, 1963) indicates that they have no requirements for unknown nutrients supplied specifically by bacteria. Furthermore the rearing of germ-free rats and mice on chemically defined, water-soluble diets (Pleasants, 1966) demonstrated no requirement of nutrients other than those already known to be required by these species. The digestive systems of germ-free and conventional animals appear more or less comparable. In fact, the activity of digestive enzymes such as disaccharidases in the mucosa of the small intestine and of proteolytic, amylolytic and lipolytic enzymes in the intestinal contents was higher in germ-free than in conventional animals (Reddy and Wostmann, 1966; Reddy et al., 1967). On the other hand, it has been shown that germ-free rabbits utilized iron more efficiently from a natural source than from the usual mineral supplement (Reddy et al., 1965) whereas conventional rabbits utilized either source equally efficiently.

It is evident that diets for germ-free animals must provide adequate levels of nutrients, notwithstanding (a) the changes produced by diet sterilization and storage, (b) the absence of microbial effects on quantity and availability of nutrients in the alimentary canal and (c) the structural and functional characteristics of an animal which has become physiologically, and perhaps genetically, adapted to the germ-free state. Since there is no evidence of qualitative differences in nutritional requirements between germ-free and conventional animals, the same general types of diets may be used for both classes but the composition may have to be modified in quantity and sometimes in the form of certain nutrients in order to compensate for the effects mentioned above.

With the availability of diets to rear gnotobiotic animals in adequate numbers, an ideal tool appears at our disposal not only to solve specific problems in which components of the intestinal microflora play an essential role but also for basic studies in immunology and nutrition. In many instances it will be necessary, however, to match the definition of the experimental animal achieved by the control of the microflora with a comparable definition of its nutritional intake and antigenic exposure.

II. Factors Affecting Nutritional Adequacy

A. STERILIZATION OF DIETS

Diets used for germ-free animals must be sterilized rigorously to ensure complete absence of viable bacteria and bacterial spores. Theoretically, the ideal method for sterilization would be one which simply eliminates microbes from a diet, leaving the nutrient composition unchanged. The milk diets used by the earliest investigations were sterilized by intermittent boiling (Nuttal and Thierfelder, 1897; Kuster, 1915; Glimstedt, 1936). In general the following methods are presently used: (i) chemical, (ii) steam, (iii) radiation and (iv) filtration of water-soluble diets.

1. CHEMICAL STERILIZATION

Although complete sterility of diets can be achieved by chemical methods, this procedure has been tested primarily on disease-free animals (Amtower and Calhoon, 1964; Charles et al., 1964; Porter and Lane-Petter, 1965). Ethylene oxide (Charles et al., 1964; Porter and Lane-Petter, 1965) and beta-propiolactone (Amtower and Calhoon, 1964) have been used for diet sterilization. A number of reports suggested harmful effects of ethylene oxide on the nutritional properties of diets. There was a complete destruction of thiamine (Hawk and Mickelson, 1955), a partial destruction of riboflavin, niacin, pyridoxine and folic acid (Bakerman et al., 1956) and of histidine and methionine (Windmueller et al., 1956). However, recent studies (Charles et al., 1964; Porter and Lane-Petter, 1965) indicated that ethylene oxide, if used correctly, appeared to be a satisfactory agent for sterilization of laboratory animal diet. The explosiveness of ethylene oxide could be eliminated if the gas was mixed with 5–7 times its own volume of CO_2, or by its use in the absence of oxygen. Charles et al. (1965) recommend a sterilization period of 8 hr at a gas concentration of 1500 ml/m³ with a chamber loaded to its capacity and the load preheated in its paper sacs to 24–27° C. The optimum sterilization conditions were reached at 26·7° C. Adequate aeration of the diet after fumigation was an important factor in reducing the toxicity of ethylene oxide. The load was rinsed with air by removing the gas at the end of the sterilization period and subsequently introducing filtered air for 5-min periods. However, from the data of these investigators it was obvious that 50% of nicotinic acid and 25% of pyridoxine in the diet had been destroyed (Charles et al., 1964; Porter and Lane-Petter, 1965). Thus, in order that the diets be complete following sterilization, an excess of certain nutrients must be added.

However, further studies of the effects of ethylene oxide and other possible sterilizing chemicals in diets and other materials are needed. Reyniers et al. (1964), in a retrospective study, have strongly implicated ethylene oxide sterilized bedding as the cause of increased cancer incidence in female germ-free mice. In addition, with beta-propiolactone sterilization, although excess of this compound hydrolyses in time to the harmless beta-hydroxy-propionic acid, the organic oxy-compounds formed during sterilization have been implicated by van Duuren et al. (1963) as possible carcinogens.

2. STEAM STERILIZATION

In most gnotobiotic operations, diets are routinely sterilized by steam under pressure. Reyniers et al. (1946) established the methods which have been used for steam-sterilization of solid diets in most subsequent work. For germ-free operations, the exposure of the diet must be sufficient to guarantee a sterile product but excessive temperatures and sterilization times and low moisture content may be deleterious, especially for certain nutrients.

In the Lobund Laboratory, University of Notre Dame, U.S.A., diets intended to be introduced into stainless steel cages are routinely auto-claved as follows: the solid diet is wrapped in cloth bags, placed in a sterile lock and a precycle vacuum of 20 in. maintained for 10 min. Then free flowing steam for 5 min is used to help flush out residual air before steam under pressure (121° C) is admitted for 25 min. The steam pressure is then brought down slowly without a post-cycle vacuum. Diets intended to be introduced into plastic isolators are autoclaved with a 5-min precycle vacuum (approximately 28 in.) followed by a 25-min period of 15 lb of steam pressure at 121° C and a 20-min post cycle vacuum at 28 in.

The sterilization of solid diets by steam presents a serious problem since it will result in loss of nutritive value of proteins particularly through its effects on lysine, methionine and cysteine (Rice and Beuk, 1953), loss of certain B vitamins, also of vitamins A and E (Zimmerman and Wostmann, 1963), formation of potential antagonists to important nutrients or metabolites (Reyniers et al., 1946) and the formation of potentially antigenic products. Theoretically some loss could also occur by leaching, but a comparison of iron and copper contents of diets before and after autoclaving showed no demonstrable loss by leaching (Reddy et al., 1965).

Recent investigations in our laboratory (Zimmerman and Wostmann, 1963) have shown that graded additions of water to the air-dried diets before autoclaving progressively improved the apparent stability of thiamine, pyridoxine and calcium pantothenate. Increases in moisture content of the diet, therefore, would not only increase the recovery of these vitamins after sterilization, but might also reduce the formation of interfering products. Thus, many of the nutrient losses resulting from steam sterilization of diets could be effectively minimized by careful control of temperature and sterilization time and of the moisture content of the diet. Furthermore, in most instances it should be possible to estimate sterilization losses and to compensate for them by enrichment of the original formula.

3. RADIATION STERILIZATION

In the course of experimental studies with germ-free animals, investigators have continuously sought means other than steam under pressure to sterilize diets containing heat-labile nutrients for germ-free animals. The use of radiation-sterilized complete diets for laboratory animals was explored by Luckey *et al.* (1955). Mice fed on such a diet for three generations demonstrated growth and reproduction equal to that of mice fed on the same diet in non-sterilized or steam-sterilized form. Chemical analyses showed a greater destruction of thiamine, riboflavin, folic acid and pantothenic acid in the steam-sterilized diet than in the irradiated diet. On the basis of the few comparative studies available, the radiation sterilization of complete diets for germ-free animals appears superior to autoclaving in terms of animal performance and recovery. Horton and Hickey (1961) raised germ-free guinea pigs on both natural-type diets and semi-synthetic type diet sterilized in a 3×10^6 V Van de Graff Electron Beam Accelerator for a total dose of 2 or 3 Mrad. The irradiated enriched natural-type diet proved to be satisfactory for conventional guinea pigs. This same diet failed to meet the nutritional requirements for the rearing of the germ-free animals and caused death in 2 weeks with greatly distended caeca. Irradiated semi-synthetic diet, however, provided substantial growth of germ-free guinea pigs, with no caecal distention (Horton and Hickey, 1961). A special advantage of the irradiated guinea pig diet was the fact that enough ascorbic acid was retained in the diet to meet the requirements of the guinea pigs. When autoclaved diets have been used, it has been necessary to sterilize ascorbic acid separately and add it to the water or the food inside the isolator. Also, irradiation of simple, dry sugars could be carried out, whereas autoclaving caramelized them.

In the studies of Coates *et al.* (1963) with germ-free chickens, an enriched natural-type diet failed to support normal growth after auto-claving, but the same diet irradiated at 5 Mrad from a ^{60}Co source supported better than conventional growth. Although vitamin K is one of the most radiosensitive nutrients and germ-free chickens are especially sensitive to a deficiency of this vitamin, the irradiated diet permitted normal clotting times. Chemical analyses of the diet showed that even the radiosensitive vitamin E and carotene suffered only 50% loss from irradiation, while thiamine was reduced only 32%. Vacuum packing of the diet before radiation reduced the vitamin E loss to a mere 10% of the original content. Growth of germ-free chickens was even better when the radiation dose was reduced from 5 to 3 Mrad, but further investigations had to be undertaken to determine if sterility could be consistently assured at this level.

It appears from the above findings that, from the standpoint of reliability, nutrient recovery and convenience, radiation sterilization might represent the best available method for sterilization of diets for germ-free animals, particularly for diets of a relatively defined character.

4. STERILIZATION BY FILTRATION

In many instances, sterilization-induced destruction and conversion will make it impossible even to estimate the actual nutrient pattern of the sterilized diet. For colony production of germ-free rats and mice, almost any balanced diet fortified to compensate for sterilization losses will produce acceptable results. However, in the fields of immunology and basic nutrition which require that uncontrolled antigenic and synergistic stimulation be kept to a minimum and that dietary intake be totally defined, the need for a more sophisticated germ-free animal arises.

Based on the pioneering work of Greenstein *et al.* (1956, 1960) the Lobund laboratory (Pleasants and Wostmann, 1962) has developed chemically defined, low molecular weight, water-soluble formulas. Such mixtures of chemically pure components can be sterilized by Millipore filtration without loss and would guarantee not only an almost absolute control of nutritional intake but also a diet of extremely low potential antigenicity. The procedure routinely used at the Lobund laboratory was developed and described in detail by Pleasants (1966; see also Chapter 3). A standard Millipore pressure filtration apparatus with a 47 mm diameter GS membrane of $0.22\,\mu$ pore size and a 42 mm diameter glass fibre pre-filter was routinely used. The delivery tube of the filtration apparatus was fitted into a Florence flask having a constricted neck. After filtration, the flask was sealed in a flame and sprayed with peracetic acid for entry into the isolator.

In some instances, sealing of glass flasks had proved especially hazardous when oils were being filtered and it also caused formation of caramelized material where diet had spattered on the flask near the place of sealing. In order to prevent this, Pleasants (1966) developed a new technique for filtering directly into the isolator through a plastic pipe cemented to the top of the entry post of the isolator itself (for details, see Chapter 3).

B. STORAGE OF DIETS

Although in many instances germ-free animals could be readily procured, acceptable growth rates and reproduction were not obtained in many laboratories. One of the major causes for this discrepancy was the destruction of certain nutrients, due to prolonged storage of diets. The pattern of these losses often resembled that observed after sterilization of diet at higher temperatures. In many cases, storage of solid diets for longer time periods presented the usual problems of thiamine, vitamin A and vitamin E destruction. In order to keep nutrient losses to a minimum, Lobund laboratory has adopted the practice of storing sterilized solid diets in the isolator at room temperature for maximally 10 days and preferably no longer than 7 days. Newton (1965) reported that at the National Institute of Health, U.S.A., sterilized diets were stored in isolators for no more than 2 weeks.

Pleasants *et al.* (1966) found that Millipore-filtered liquid diets tended to brown rapidly at isolator temperatures, thereby losing some of their chemical definition and conceivably developing antigenicity. The colour changes were not noticeable during two weeks of storage at room temperature when liquid diets were filtered and stored in two separate solutions, the first containing amino acids and vitamins and the second containing sugars and minerals. These two solutions were combined just prior to feeding (see also Section III A2).

C. INFLUENCE OF INTESTINAL MICROFLORA ON NUTRIENT REQUIREMENT

The concept that the normal intestinal microflora of the non-ruminant mammal might affect its nutrition and health is, of course, not new but the influence of nutrient production by microflora is considered more pronounced in the coprophagous species such as rodents and rabbits. The complete prevention of coprophagy created major changes in the requirements for vitamin K, biotin, thiamine, pyridoxine, pantothenic acid, riboflavin, vitamin B_{12}, essential fatty acids and possibly folic acid (Barnes *et al.*, 1963). Many of the flora-produced micronutrients are not effectively absorbed in the absence of coprophagy. In order to be absorbed effectively these micronutrients, which are excreted in the faeces, must be returned to the upper intestinal tract through the practice of coprophagy (Barnes *et al.*, 1963). Folic acid, on the other hand, was not only supplied to the animals through recirculation of intestinal contents, but this microbial product was also in part available via direct absorption (Daft *et al.*, 1963). Therefore, it is essential to provide high levels of micronutrients in germ-free animal diets as a safety factor until exact requirements can be worked out.

It is also true, on the other hand, that the intestinal microflora might compete with the host organism for nutrients such as amino acids and vitamins, thereby aggravating nutritional deficiencies and increasing nutritional requirements. Levenson and Tennant (1963) reported that germ-free guinea pigs fed a scorbutogenic diet developed scurvy much more slowly than the conventional animals. Ascorbic acid was probably utilized by the intestinal bacteria and thereby lost to the host. In addition, certain elements of the normal intestinal flora have been shown to depress growth of mice on low protein diets (Dubos and Schaedler, 1960).

The intestinal flora may also interfere with the absorption of nutrients by causing an increase in thickness of the intestinal mucosa and mild forms of toxaemia which decrease the efficiency of food utilization (Dubos *et al.*, 1963). Thus, many additional characteristics produced by the absence of a microflora in the gut tract, e.g. a more positive oxidation-reduction potential (Wostmann and Bruckner-Kardoss, 1966), a reduction of mucosal surface area (Gordon and Bruckner-Kardoss, 1961), changes in the pattern of development of mucosa (Abrams *et al.*, 1963), a reduction in lamina propria connective tissue (Gordon and Bruckner-Kardoss, 1961) and a lowered tissue hydration (Gordon, 1959) may influence the efficiency of food utilization.

D. INFLUENCE OF INTESTINAL MICROFLORA ON THE AVAILABILITY OF NUTRIENTS

Germ-free rabbits in our laboratory demonstrated bone fragility, hind leg paralysis, irregular pregnancies and low haemoglobin and plasma iron levels, whereas none of these symptoms occurred in their conventional counterparts maintained with the same sterilized diets L-473E5 or L-477, containing recommended levels of nutrients (Reddy *et al.*, 1965). Either conventionalization of these germ-free rabbits or transfer to a sterilized diet L-478 which contained more iron from natural ingredients (soya bean meal) but less total amount of this mineral alleviated all symptoms of iron-deficiency anaemia in 4 weeks. Since the amount of iron present in diets L-473E5 and L-477 was adequate for conventional rabbits, its availability must have been influenced by the presence of intestinal microorganisms.

On the other hand, certain types of diets for rats and mice appeared to result in excessive absorption of calcium by germ-free but not by conventional animals. Urinary calculi and elevated urinary calcium were observed in germ-free rats on diets which produced no such symptoms in conventional controls (Gustafsson and Norman, 1962). Soft tissue calcification and urinary calculi observed in female germ-free C3H mice were associated with a particular type of diet which, however, did not adversely affect the conventional control animals (Reyniers and Sacksteder, 1958). Other types of diet did not produce this effect, although they were as high in calcium content, thus indicating that other factors besides dietary calcium are involved in this phenomenon.

In conclusion, the above examples indicate that the testing of sterilized diets with conventional animals cannot assure their adequacy for germ-free animals and that the form in which a nutrient is provided might be of more critical importance for the germ-free than for the conventional animal. The germ-free animal may be successfully raised by the application of general nutritional principles, provided the experimenter is aware of the ways in which diet sterilization, diet storage and presence or absence of microflora could influence the dietary requirements.

III. Composition of Diets for Germ-free Animals

A. RATS AND MICE

During the initial period of germ-free operations when colonies were being derived by surgical operation and hand-feeding, diets for germ-free mammals fell into two groups: milk formulas to cover the period until weaning and solid diets for the post-weaning period. Since colonies of germ-free rats and mice are now in existence at many research institutes and commercial undertakings, it is no longer necessary to obtain germ-free rats and mice by surgical operation and subsequent hand-feeding. Hand-feeding of germ-free rats and mice is now carried out mostly for special purposes and its problems and progress are discussed in detail in Chapter 3.

1. SOLID DIETS

With the impressive progress made so far in methodology and sterilization techniques, almost any balanced diet fortified with vitamins to compensate for sterilization losses has sustained adequate growth and reproduction in germ-free rats and mice. An increased knowledge of nutritional requirements, especially of germ-free animals, and careful breeding techniques made these results possible.

Practical type and semi-synthetic type diets developed by Gustafsson (1959), Luckey (1963), Wostmann (1959) and several commercial producers were found to give satisfactory results for colony production of germ-free rats and mice. Formula L-356, a semi-synthetic diet based on rice flour, casein, yeast extract, liver powder, corn oil, salts and added vitamins (Luckey, 1963) was essentially an experimental one which had a comparatively low antigenicity. Diet L-462, a more practical type, was composed of wheat flour, corn meal, whole milk powder, milk protein, alfalfa and liver with vitamins and salts added (Wostmann, 1959). At the Lobund laboratory, diets L-356 and L-462 have been used successfully to raise many generations of germ-free rats and mice without noticeable deficiency symptoms or other abnormalities (Wostmann, 1959).

It is of obvious importance that the investigator working with germ-free animals be able to exert the amount of dietary control which is required by the experiment and which is in keeping with the sophistication and relatively high cost of this experimental animal. With these scientific and economic considerations in mind, the Lobund laboratory has developed two new series of solid diets for germ-free rats and mice.

TABLE I

Composition of solid diets for germ-free rats

Ingredient	Diets	
	L–474 E12	L–484
	%	%
Casein, vitamin-free	24·0	—
Corn, ground	—	58·96
Soya bean meal, 50%	—	30·0
Alfalfa meal, 17%	—	3·5
Cellophane spangles	5·0	—
Rice starch	60·4	—
Corn oil	3·0	3·0
Lysine, feed grade	—	0·5
Methionine, feed grade	0·3	0·5
Inositol	0·1	—
Butylated hydroxytoluene	—	0·0125
Ca CO_3	1·7	0·5
K_2HPO_4	1·0	—
Na_2HPO_4	1·0	—
Ca $HPO_4 \cdot 2H_2O$	—	1·0
Salts–27[a]	1·0	—
NaCl, iodized	—	1·0
Trace mineral mixture[b]	—	0·025
Ladek–55[c]	2·0	—
B-vitamin mixture–75[d]	0·5	—
Vitamin premix[e]	—	1·0

[a] Salts–27 contained: (in mg) NaCl, iodized, 515; $MgSO_4$, 400; Fe $(C_6H_5O_7)_2$, 60; $MnCO_3$, 20; CuO, 2·5; ZnO, 2·5; $CoCl_2 \cdot 6H_2O$, 0·05; NaF, 0·01; MoO_3, 0·005; KBr, 0·01; Na_2SeO_3, 0·01.

[b] Trace mineral mixture contained: (mg/100 g diet) Mn, 6·58; Fe, 2·74; Cu, 0·22; Zn, 1·31; I, 0·13; Co, 0·066.

[c] Ladek–55 contained: dl-α-tocopheryl acetate, 10 mg; vitamin A concentrate, natural ester form, 1600 IU; vitamin D_3, 100 IU; vitamin E, mixed tocopherols, 37·5 mg; vitamin K_3, 10 mg; corn oil, 2 g.

[d] B vitamin mixture–75 contained: (in mg) thiamine, 6; riboflavin, 3; nicotinamide, 5; calcium pantothenate, 30; choline chloride, 200; pyridoxine hydrochloride, 2; pyridoxamine dihydrochloride, 0·4; biotin, 0·1; folic acid, 1; para-aminobenzoic acid, 5; rice starch carrier, 217·5.

[e] Vitamin premix contained: (mg/100 g diet) vitamin A, 2632 IU; vitamin D_3, 101; vitamin E, 21·9; vitamin K_3, 8·77; riboflavin, 3·07; pantothenic acid, 28·51; niacin, 6·58; choline chloride, 197; vitamin B_{12}, 0·44; thiamine, 6·58; pyridoxine hydrochloride, 2·19; folic acid, 1·10; para-aminobenzoic acid, 5·04.

Formula L-356, which proved its value during many years of use, was simplified and further defined. This resulted in semi-synthetic diet L-474 (Table I) based on extracted casein and rice starch and containing corn oil or ethyl linoleate with added methionine, vitamins and minerals (Wostmann and Kellogg, 1967). Both germ-free and conventional rats fed on this diet showed normal and comparable weight gain. Germ-free rats have been maintained on this formula for about a year without signs of nutritional deficiency. The L-474 formulation has been used routinely with good results in experimental work which required more extensive nutritional control.

Although diet L-462 supported colony production of germ-free rats and mice for many generations at the Lobund laboratory, the relatively high cost of this diet led to the development of a low cost diet L-484 (Table I). The formula was based on maize, soya bean meal, alfalfa meal and corn oil with added feed-grade amino acids, vitamins and salt mixture. These ingredients are readily available, uniform, inexpensive and result in a diet that can be commercially produced for less than $0.15 per pound (Kellogg, 1966). In a study conducted at the Lobund laboratory on the performance of germ-free rats and mice maintained on this diet, Kellogg (1966 and unpublished observations) reported that growth of weanlings was satisfactory and a fourth generation was successfully reared on this diet.

2. CHEMICALLY DEFINED, WATER-SOLUBLE DIETS

Hand-feeding of surgically-derived germ-free rats and mice required a liquid formulation to replace the mother's milk. Encouraging results had been obtained by using diet L-449C (Wostmann, 1959) and later diet formulations L-460C (Pleasants, 1959) which resembled the composition of natural rat milk. It was shown that 85% of young rats started were weaned and the only deaths during the feeding period were due to accidental causes (Pleasants, 1959). However, the protein concentration in these milk formulas could not be raised to more than 6% without encountering digestive problems manifested by substantial amounts of undigested protein throughout the entire length of the small intestine. This was attributed to the impaired proteolytic capacity of the hand-fed newborn animals (Wostmann, 1959). Water-soluble, chemically defined diets which had been formulated at the U.S. National Institutes of Health by Greenstein et al. (1956; 1960) on the basis of amino acids and glucose, appeared to offer a potential solution since the nutrients are pre-digested.

Another potential value of this diet results from its non-antigenic character. Dietary composition can play a major role in determining the state of the animal's antimicrobial defence potential. In germ-free animals the level of immune globulins was found to be diet-dependent

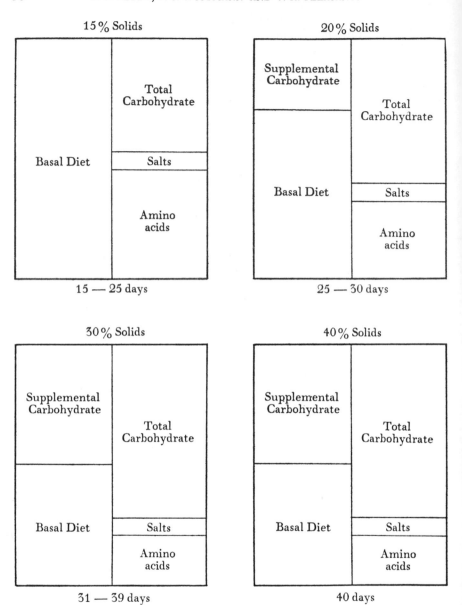

FIG. 1 Gradual increase in solids content of liquid diet and in percentage of total solids supplied by carbohydrate in diet fed to germ-free rats weaned directly from mother's milk to water-soluble diet (Pleasants, 1966).

(Wostmann, 1961; Olson and Wostmann, 1964). The severity of the syndrome resulting from association of germ-free rats with *Salmonella typhimurium* depended on the diet to the extent that animals fed diet L-462 consistently suffered 25% mortality whereas those fed diets L-481 or 5010C (Purina Lab Chow) showed a moderate stress and loss of weight but no mortality.

Furthermore, to obtain absolute dietary control it is necessary to feed a chemically defined diet from birth. To avoid all exogenous antigenicity in order to provide a baseline for immunological studies, this diet must not contain potentially antigenic macromolecules. Both desiderata demand a chemically defined, low molecular weight diet and method of sterilization which introduces no changes. So far this can be found only in sterilization by filtration of a defined, water-soluble formula.

The above goals necessitated a number of basic changes in the original formulation given by the Greenstein group (1956). Chemical definition of the formula required the purest form of unsaturated fatty acid as the lipid component of the diet, but emulsifying agents, such as Tween 80, had to be avoided because of possible stimulation of the reticulo-endo-thelial system (Mori and Kato, 1959) and alteration of the activity of certain enzymes (Holt *et al.*, 1963) by these compounds. In order to prevent the browning of Millipore-filtered diet during the usual storage of up to one week at room (isolator) temperature, this diet was prepared, filtered and stored in two separate and relatively stable solutions, one containing dextrose and minerals and the other containing amino acids and vitamins. These two solutions were mixed in the isolator before feeding. Diet and water were available from overhead bottles with punctured caps. Ethyl linoleate or corn oil containing the fat soluble vitamins was fed separately with a syringe once daily. In some cases the animals consumed these daily supplements of fat from a small dish fastened to the side of the cage. The formulation L-479E9 presently in use at the Lobund laboratory is given in Table II (Pleasants, 1966).

A special regime (Fig. 1) was followed in order to allow weanling germ-free rats and mice to adjust more easily to the low molecular weight, water-soluble formulation with its necessarily relatively high osmolarity (Pleasants, 1966). This schedule allowed for a gradual increase in carbohydrate and in total solid concentration during the first 4–5 weeks of life. Germ-free rats appeared grossly and histologically normal and showed no evidence of specific deficiencies. After a short lag period caused by the necessary adjustment needed for absorbing and utilizing a low molecular weight formula of high osmolarity, germ-free rats fed on this water-soluble diet showed weight gains comparable to those of rats fed on solid diets (Pleasants, 1966).

Germ-free mice fed on this liquid diet from weaning showed normal growth and normal reproduction, making possible the start of a reproducing germ-free mouse colony maintained under defined conditions

TABLE II

Composition of chemically-defined, water-soluble diet (L–479 E9) for germ-free rats and mice

Ingredients	g/100 g solids	Ingredients	g/100 g solids
	amino acids–vitamins		
L-lysine HCl	1·25	L-asparagine	1·20
L-histidine HCl·H$_2$O	0·55	L-proline	3·00
L-tryptophan	0·40	Monosodium l-glutamate	6·00
L-phenylalanine	0·90	Glycine	0·50
L-isoleucine	0·50	L-serine	1·55
L-leucine	0·80	L-alanine	0·75
L-threonine	0·50	Tyrosine-ethyl ester HCl	2·00
L-methionine	0·85		
L-valine	0·70		
L-arginine	0·75		
B vitamin mix 111 E2[a]	0·065	NaCl + KI	0·38
Choline Cl	0·25	K(C$_2$H$_3$O$_2$)	1·06
Ladek–62[b]			

	sugars–minerals		
			(mg/100 g solids)
Dextrose	20·5	Mn (Ac)$_2$·4H$_2$O	26·0
Ca fructose (PO$_4$)$_2$	5·0	ZnSO$_4$·H$_2$O	5·5
Mg fructose (PO$_4$)$_2$	0·5	Cu (Ac)$_2$·H$_2$O	2·5
Ferrous gluconate	0·035	Co (Ac)$_2$·4H$_2$O	0·9
Supplemental carbohydrate	50·0	(NH$_4$)$_6$ Mo$_7$O$_{24}$·4H$_2$O	0·6
		Na$_2$SeO$_3$	0·011
		Cr (Ac)$_3$·H$_2$O	0·48

[a] B vitamin mix–11E2 contained: (mg/100 g diet) thiamine HCl, 0·50; riboflavin, 0·75; pyridoxine HCl, 0·63; niacin, 3·75; inositol, 25; calcium pantothenate, 5; p-aminobenzoic acid, 30; biotin, 0·10; folic acid, 0·15; cyanocobalamin, 0·03.

[b] Ladek–62 was administered orally (200 mg/day/animal). Daily supplement contained: (in mg) dl-α-tocopherol, 2; dl-α-tocopheryl acetate, 4; vitamin A palmitate, 0·33; vitamin K, 0·55; vitamin D$_3$, 0·35 μg.

(Pleasants et al., 1964). So far germ-free CFW mice maintained on this formulation have reproduced into the third generation. During the last week of lactation it is apparently difficult for the mother to satisfy nutritional requirement of herself and her offspring. This resulted in loss of weight of the mother as well as in subnormal weaning weights of the young. However, the deficit was quickly overcome after weaning.

Germ-free rats have been successfully maintained on these water-soluble, chemically defined, low molecular weight formulas from birth to adulthood. The animals appeared grossly normal (Pleasants, 1966). When given this diet with varying proportions of total solids, fats and carbohydrates as shown in Fig. 2, at half hour intervals, newborn germ-free

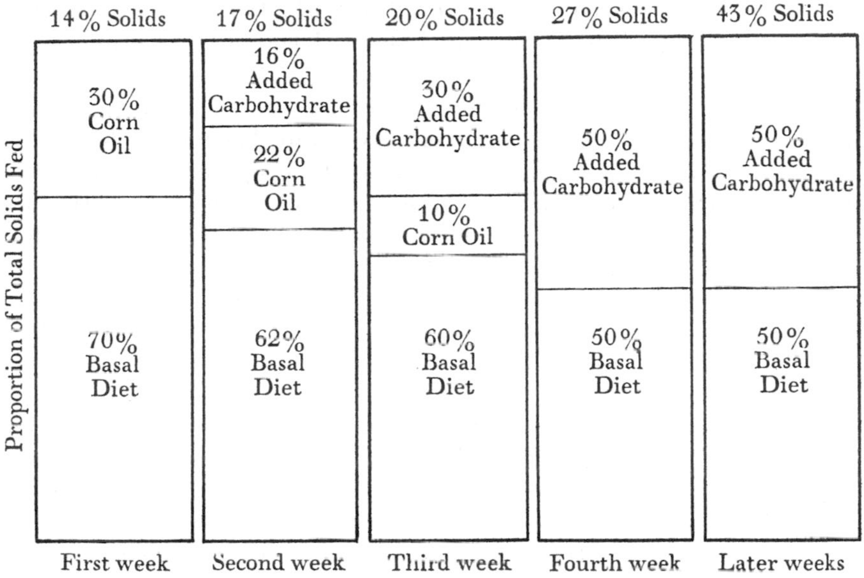

FIG. 2 Changing proportions of calorie sources during hand feeding of water-soluble diet (Pleasants, 1966).

rats showed a steady gain in weight (Pleasants, 1966). The more frequent feeding of small amounts of diet was found to hasten stomach emptying and reduce intestinal distention. Although infant germ-free mice were not successfully hand reared on water-soluble diets, one mouse survived as long as 12 days. The high mortality rate of hand-fed mice appeared to be caused mainly by the mechanical difficulties of holding and feeding such delicate animals weighing 1·4 to 1·8 g at birth (Pleasants, 1966). This is amply illustrated by the fact that mice are always much more difficult to hand-rear than rats, even on milk formula (Pleasants, 1959).

The facts that three successive generations of germ-free mice have been obtained with these diets, and that germ-free rats can be raised from birth to adulthood, suggest that the present defined formulation covers all nutritional requirements of germ-free rats and mice. The statement seems warranted that all nutritional requirements of these rodents can be met with presently known, chemically defined low molecular weight nutrients.

B. GUINEA PIGS AND RABBITS

In recent times, germ-free guinea pigs have been reared in a number of laboratories (Glimstedt, 1936; Horton and Hickey, 1961; Miyakawa, 1955; Newton and DeWitt, 1961; Phillips *et al.*, 1959) but reproduction had only been reported from the Lobund laboratory (Wostmann and Pleasants, 1959) by animals reared on a diet slightly modified from that of Phillips *et al.* (1959). Reproduction on this diet was only sporadic and the animals showed extreme distention of the caecum. The results obtained by Horton and Hickey (1961) with an irradiated semi-purified diet indicated that the caecal size of germ-free guinea pigs could be brought within the conventional range by means of dietary changes. At the Lobund laboratory the successful germ-free rat diet L-462 (Wostmann, 1959) was then modified to meet the requirements of the germ-free guinea pig. The major modifications were: addition of magnesium and potassium acetates to satisfy higher need for alkali in the guinea pig, deletion of skim milk powder, addition of rolled oats, reduction of iron and copper levels and addition of 25% water before autoclaving (Pleasants *et al.*, 1967). The resulting diet L-477 (Table III) and the supplement of 20 mg ascorbic acid and 1 mg thiamine per guinea pig per day via the drinking water maintained a nearly normal standard of growth, normal gross morphology and good reproduction through four generations of germ-free guinea pigs (Pleasants *et al.*, 1967).

Germ-free rabbits showed poor post-weaning growth and no reproduction when fed on modified commercial type diets (Wostmann and Pleasants, 1959). Diet L-477 which supported good growth and reproduction in germ-free guinea pigs did not meet the requirements of the germ-free rabbit. Bone fragility, irregular pregnancies and iron-deficiency anaemia were observed although there was an ample supply of iron in the diet. It was postulated that the availability of iron to germ-free rabbits was reduced by the absence of intestinal microbiota, but that the resulting anaemia might be prevented by providing iron in a form whose availability was not seriously affected by the germ-free condition (Reddy *et al.*, 1965). After a series of earlier modifications, diet L-478 (Table III) was formulated which incorporated both soya bean meal and a mineral mixture in which $MgSO_4$ was replaced largely by $MgCO_3$. Germ-free rabbits maintained on this diet appeared normal and healthy and showed normal growth and reproduction (Reddy *et al.*, 1965).

1. FEEDING OF SURGICALLY-DERIVED GERM-FREE GUINEA PIGS AND RABBITS

Germ-free guinea pigs, which need no milk formula and can eat independently after birth, have been weaned successfully either on diets L-445 (Phillips *et al.*, 1959) or L-477 (Pleasants *et al.*, 1964). During the

TABLE III

Composition of solid diets for germ-free rabbits and guinea pigs

Ingredient	Diets	
	L-477	L-478
	%	%
Wheat, ground	32	20
Corn, ground, yellow	14·55	15
Oats, rolled	22	18
Soya bean meal	—	25
Lactalbumin	10	5
Casein	7·8	—
Alfalfa, dehydrated	2	5
Liver, desiccated	2	2
Corn oil	3	3
$CaCO_3$	1·1	0·75
$CaHPO_4 \cdot 2H_2O$	0·4	1
$Mg(C_2H_3O_2)_2 \cdot 4H_2O$	0·95	0·75
CH_3COOK	0·9	1·2
Salts–24[a]	1	—
Salts–26[b]	—	1
Ladek–3 (fat soluble vitamins)[c]	2	2
B vitamin mixture–75[d]	0·25	0·25
Inositol	0·05	0·05

[a] Salts–24 contained: (in mg) NaCl (iodized), 530; $MgSO_4$, 400; $FeC_6H_5O_7 \cdot 5H_2O$, 60; CuO, 2·5; $MnCO_3$, 4·2; ZnO, 2·5; (in μg) MoO_3, 15; NaF, 22; $CoCl_2 \cdot 6H_2O$, 400; KBr, 10; Na_2SeO_3, 10.

[b] Salts–26 contained: (in mg) NaCl (iodized), 530; $MgSO_4$, 100; $MgCO_3$, 300; $FeC_6H_5O_7 \cdot 5H_2O$, 80; CuO, 3·8; $MnCO_3$, 4·2; ZnO, 2·5; (in μg) MoO_3, 15; NaF, 22; $CoCl_2 \cdot 6H_2O$, 400; KBr, 10; Na_2SeO_3, 10.

[c] Ladek–3 contained: *dl*-α-tocopheryl acetate, 10 mg; vitamin A concentrate, natural ester form, 800 IU; vitamin D, 100 IU; vitamin E, mixed tocopherols, 150 mg; vitamin K_3, 10 mg; corn oil carrier, 2 g.

[d] B vitamin mixture–75: see Table I for composition.

first two weeks after surgical delivery, the diet was sterilized in pans with 5 times its weight of water and given with a separately autoclaved supplement of 1% ascorbic acid and 0·05% thiamine HCl. After two weeks of age, the guinea pigs were offered the same diet L-477 (Table III) with only 25% water added before autoclaving.

Earlier methods of hand-rearing germ-free rabbits had provided only 23% survival to weaning and very limited postweaning survival (Pleasants, 1959). Most of the infant rabbits were lost by the aspiration of regurgitated diet resulting either from over-filling of the stomach from too rapid feeding, or from an excessively fluid consistency of the curd formed in the stomach. The diet used tended to coagulate during autoclaving. Pleasants et al. (1964) improved both the hand-rearing technique

TABLE IV

Composition of milk diet L-466 E3 for surgically-derived germ-free rabbits

Ingredients			
Combined in milk ampoule		To be ampouled and autoclaved separately	
Centrifuged skim milk		B vitamin mix–103[b]	3 ml
concentrate	43·0 ml	Salts–15A[c]	1 ml
Coffee cream (18% butterfat)	57·0 ml	Salts–15B[d]	1 ml
Vi-syneral, fortified[a]	0·2 ml	$CaCO_3$, 8% suspension in water	3 ml
DL-Methionine	0·14 g	Calcium ascorbate, 1% in	
L-tryptophan	0·05 g	tripled distilled water	5 ml

[a] To Vi-syneral are added (per 0·2 ml): vitamin K_3, 0·33 mg; mixed tocopherols, 3·3 mg.

[b] B vitamin mix–103 in 3 ml triple distilled water contained: (in mg) thiamine HCl, 3; riboflavin, 1; pyridoxine HCl, 1; calcium pantothenate, 12·5; niacinamide, 2·5; folic acid, 0·25; biotin, 0·025; cyanocobalamin, 0·025; choline dihydrogen citrate, 100; inositol, 100.

[c] Salts–15A in 1 ml distilled H_2O contained: (in mg) Na_2HPO_4, 180; KH_2PO_4, 176; KI, 0·4.

[d] Salts–15B in 1 ml distilled H_2O contained: (in mg) ferric ammonium citrate, 24; $MgSO_4$, 14; $MnCl_2 \cdot 4H_2O$, 4; $ZnSO_4 \cdot H_2O$, 5·2; $CuCl_2$, 2·4; $CoCl_2 \cdot 6H_2O$, 0·8.

and the milk formula and attained 75% survival through weaning. The milk formula L-466 E3 (Table IV) contained about 13% milk protein concentrated from skim milk by centrifugation to minimize changes in the protein and thus make it less susceptible to coagulation during sterilization (Pleasants et al., 1964). The sterilized formula was fed to surgically derived germ-free rabbits which were subsequently weaned at the age of 21 days on to solid diet L-478.

C. CHICKENS AND TURKEYS

Practical type diets developed at Lobund laboratory L-124 and L-289 (Luckey, 1963) and at The National Institute for Research in Dairying,

Shinfield, Reading, U.K. (Coates *et al.*, 1963) and diets based on casein-starch or soya-corn (Forbes and Park, 1959) have been used successfully with germ-free chickens. Better growth was obtained when potassium and calcium were provided in the diet in the form of potassium mono-phosphate and calcium carbonate rather than as calcium phosphate (Forbes and Park, 1959). For the last 10 years, germ-free chickens have been maintained at the Lobund laboratory on diet L-289F (Table V), which supported normal growth well into maturity. All birds appeared grossly normal and no specific signs of deficiencies have been noted.

Germ-free turkeys have been raised successfully on diet L-318 based on a soya-corn mixture (Luckey *et al.*, 1960).

TABLE V

Composition of diet L-289F for germ-free chickens

Ingredients	g/100 g diet
Corn, ground, yellow	48·0
Soya bean meal	23·3
Alfalfa meal	2·79
Wheat middlings	9·3
Fish meal	2·32
Meat scrap powder	2·32
Casein, crude, 30 mesh	7·0
NaCl, iodized	0·47
$CaCO_3$	1·685
$CaHPO_4 \cdot 2H_2O$	1·5
$MnSO_4 \cdot H_2O$	0·025
Fat soluble vitamin mixture in corn oil[a]	1·0
B and C vitamin mixture[b]	0·2

[a] Fat soluble vitamin mixture in 1 g corn oil contained: vitamin A, 1400 IU; vitamin D, 300 IU; vitamin E (mixed tocopherols, 253·3 IU/g), 75 mg; vitamin K_3, 5 mg.

[b] B and C vitamin mixture contained: (in mg) ascorbic acid, 20; thiamine HCl, 2·18; inositol, 20; riboflavin, 0·72; pyridoxine HCl, 0·72; calcium pantothenate, 3·09; niacin, 1·6; nicotinamide, 2; biotin, 0·01; folic acid, 0·4; cyanocobalamin, 0·0024; choline chloride, 46·2; corn starch carrier to 200.

D. SHEEP AND GOATS

Surgically derived germ-free sheep and goats have been maintained on steam-sterilized fortified milk formulae (Table VI) for several months (Smith and Trexler, 1960; Smith, 1966). In animals receiving fortified milk formula at 15% of their body weight, growth rate was generally adequate. However, the heart and kidneys of the germ-free animals showed degenerative conditions associated with hypervitaminosis D. Withdrawal of the vitamin D supplemented milk from the diet alleviated these symptoms (Smith, 1966). Smith (1966) reported that germ-free goats maintained on these formulations showed hyperplasia and emphysema of the lungs, presumably indicating an allergic reaction to bovine milk protein. However, this problem could in theory be overcome by establishing a homologous system by rearing germ-free goats with sterilized goat's milk (Smith, 1966).

TABLE VI

Composition of milk formulae for germ-free sheep and goats

Ingredients	Diets			
	A	B	C	D
			ml	
Cow's milk (homogenized vitamin D)	965	—	—	—
Cow's milk (skim)	—	965	—	—
Cow's milk (skim and homogenized 1:1)	—	—	965	985
B vitamin mix–103[a]	15	15	15	15
Salts–15, solutions A and B[b]	20	20	20	—
dl-α-tocopheryl acetate	—	—	50 IU	50 IU

[a] B vitamin mix–103 contained: (mg/1·5 ml solution) thiamine HCl, 3; riboflavin, 1; pyridoxine HCl, 1; niacinamide, 2·5; calcium pantothenate, 12·5; choline dihydrogen citrate, 100; biotin, 0·025; folic acid, 0·25; vitamin B_{12}, 0·025; inositol, 100.

[b] Salts–15, Solution A contained: (mg/ml solution) KH_2PO_4, 176; Na_2HPO_4, 180; KI, 0·4. Salts–15, Solution B contained: (mg/ml solution) $MgSO_4$, 14; $MnCl_2·4H_2O$, 4; ferric ammonium citrate, 24; $CuCl_2$, 2·4; $ZnSO_4·H_2O$, 5·2; $CoCl_2·6H_2O$, 0·8.

E. PIGS

Attempts to rear germ-free pigs have been reported by many investigators (Landy and Sandberg, 1961; Meyer et al., 1963; Waxler et al., 1966; Whitehair and Waxler, 1963). The diet used by Meyer et al. (1963) consisted of autoclaved homogenized cow's milk fortified with 5 ml of mineral supplement solution ($FeSO_4·7H_2O$, 50 g; $CuSO_4.5H_2O$, 4 g; $MnCl_2·4H_2O$, 3·6 g; KI, 0·26 g per l of tap water), 20 g dextrose, 1 egg, 2 ml codliver oil, 500 units USP vitamin D, 1 g melted agar and 0·1 g yeast extract per quart of milk. Although these investigators did not report on the growth rate of the newborn germ-free pigs maintained on this formula, Whitehair and Waxler (1963) and Waxler et al. (1966) have reported that a diet based on cow's milk was adequate to support life in the young germ-free pig, but the slow growth rates emphasized the need for a diet based on sow's milk. Preliminary results reported by Waxler *et al.* (1966) indicate that a commercially prepared, sterilized diet (SPF Lac, Borden Company, N.Y., U.S.A.) was nutritionally adequate for young pigs kept for a relatively short time under gnotobiotic conditions.

F. CATS AND DOGS

Rohovsky *et al.* (1966) reported that cats were maintained under germ-free conditions for 245 days. Kittens fed the sterile liquid diet, Varamel (Baker Milk Co., Cleveland, Ohio, U.S.A.), demonstrated a progressive

TABLE VII

Composition of amino acid-B vitamin supplement added to
liquid diet (Varamel) for germ-free kittens[a]

Ingredients	g/l		g/l
		Amino acids	
L-arginine HCl	3·1	L valine	6·4
L-histidine HCl·H₂O	4·2	L-arginine	3·10
L-isoleucine	3·3	L-aspartic acid	6·75
L-leucine	12·1	L-cysteine ethyl ester HCl	0·62
L-lysine HCl	6·3	Sodium l-glutamate	28·20
L-methionine	7·1	Glycine	2·05
L-phenylalanine	5·4	L-proline	12·70
L-threonine	4·7	L-serine	6·60
L-tryptophan	8·0	L-tyrosine ethyl ester HCl	6·40
		B vitamins	
Pyridoxine HCl	4·00	Riboflavin	0·10
Niacine	4·00	p-aminobenzoic acid	2·00
Calcium pantothenate	4·00	Folic acid	25·00
Thiamine HCl	1·00	Biotin	0·002

[a] The complete diet consisted of a milk formula, Varamel (Baker Milk Co., Cleveland, Ohio, U.S.A.) to which was added the above liquid vitamin and amino acid supplement in the ratio of 1 part supplement to 30 parts of formula.

alopecia and a higher rate of mortality. Addition of Millipore-filtered multiple B vitamins and an amino acid mixture in the ratio of 1 part supplement to 30 parts of liquid diet alleviated all symptoms (Rohovsky *et al.*, 1966). The growth rate of germ-free kittens fed this fortified milk formula (Table VII) was comparable to that of suckling kittens (Rohovsky *et al.*, 1966). At about 4 weeks of age, the liquid diet was gradually replaced by pre-sterilized Hills prescription solid diet C/D (Hill Packing Company, Topeka, Kansas, U.S.A.).

Germ-free dogs have been raised for surgical research by Heneghan and associates at Louisiana State University Medical School, U.S.A. Within 4–6 hr after delivery, the newborn puppies were fed autoclaved Esbilac, a commercially prepared bitch's milk replacement (2 water:1 formula) fortified with 1 ml of Betalin complex (B vitamins with ascorbic acid) per litre of formula (Heneghan et al., 1966). A 2·5% methionine solution was also added to the formula in quantities to provide each puppy 50 mg methionine/day. After weaning, the puppies were maintained on steam sterilized Purina Dog Chow (Ralston Purina Company, St. Louis, Mo., U.S.A.) and water containing 1 ml vitamin B complex per 2 l of water (Heneghan et al., 1966). Germ-free dogs maintained on these diets showed acceptable growth.

G. MONKEYS

Wolfe et al. (1966) described hand-rearing of germ-free monkeys. Starting 6–8 hr after birth, the infant monkeys were given 10% glucose every 2 hr. The diet was changed to 2:1 mixture of a liquid human infant formula (Varamel, The Baker Milk Co., Cleveland, Ohio, U.S.A.) and 5% glucose at 18–20 hr after birth. Varamel alone was given from 2–3 weeks of age until solid food, Hill's prescription diet C/D (Hill Packing Co., Topeka, Kansas, U.S.A.) was added to the diet at 4 months of age. The germ-free monkeys grew as well or faster than those raised in conventional laboratory environment but were similar in appearance and behaviour (Wolfe et al., 1966).

IV. Conclusions

The development of adequate diets for germ-free animals has been approached both systematically and empirically. At present it can be concluded that various solid diets, both practical and experimental, appear to satisfy adequately the qualitative dietary requirements of germ-free rats and mice. Nutritional adequacy of the water-soluble, low antigenic diets for germ-free rats and mice was demonstrated by the facts that germ-free rats have been maintained on these diets from birth to adulthood and germ-free mice from weaning to adulthood, and that germ-free mice were able to reproduce into the third generation. These results suggest that in the case of germ-free rats and mice, nutritional requirements can be met with the presently known chemically pure nutrients.

Practical type diets, which appear nutritionally complete, are available for germ-free rabbits, guinea pigs and chickens. The diets were adequate for starting new colonies of germ-free rabbits and guinea pigs. Although diets for germ-free sheep, goats, pigs, cats, dogs and monkeys have been used for short term experiments without major problems, the development of diets for their long-term rearing is still in the experimental stage.

REFERENCES

Abrams, G. D., Bauer, H. and Sprinz, H. (1963). *Lab. Invest.* **12**, 355.

Amtower, W. C. and Calhoon, J. R. (1964). *Lab. Anim. Care* **14**, 382.

Bakerman, H., Romine, M., Schricker, J. A., Tawahashi, S. M. and Mickelson, O. (1956). *J. agric. Fd Chem.* **4**, 956.

Barnes, R. H., Fiala, G. and Kwong, E. (1963). *Fedn Proc. Fedn Am. Socs exp. Biol.* **22**, 125.

Charles, R. T., Stevenson, D. E. and Walker, A. I. T. (1965). *Lab. Anim. Care* **15**, 321.

Coates, M. E., Fuller, R., Harrison, G. F., Lev, M. and Suffolk, S. F. (1963). *Br. J. Nutr.* **17**, 141.

Daft, F. S., McDaniel, E. G., Hermans, L. G., Romine, M. K. and Hegner, J. R. (1963). *Fedn Proc. Fedn Am. Socs exp. Biol.* **22**, 129.

Dubos, R. and Schaedler, R. W. (1960). *J. exp. Med.* **111**, 407.

Dubos, R., Schaedler, R. W. and Costello, R. (1963). *Fedn Proc. Fedn Am. Socs exp. Biol.* **22**, 1322.

Forbes, M. and Park, J. T. (1959). *J. Nutr.* **67**, 69.

Glimstedt, G. (1936). *Acta path. microbiol. scand.*, Suppl. No. **30**, 1.

Gordon, H. A. (1959). *Ann. N.Y. Acad. Sci.* **78**, 208.

Gordon, H. A. and Bruckner-Kardoss, E. (1961). *Am. J. Physiol.* **201**, 175.

Gordon, H. A. and Bruckner-Kardoss, E. (1961a). *Acta anat.* **44**, 210.

Greenstein, J. P., Birnbaum, S. M. and Winitz, M. (1956). *Archs Biochem. Biophys.* **63**, 266.

Greenstein, J. P., Otey, M. C., Birnbaum, S. M. and Winitz, M. (1960). *J. natn. Cancer Inst.* **24**, 211.

Gustafsson, B. E. (1959). *Ann. N.Y. Acad. Sci.* **78**, 17.

Gustafsson, B. E. and Norman, A. (1962). *J. exp. Med.* **116**, 273.

Hawk, E. A. and Mickelson, O. (1955). *Science, N.Y.* **121**, 442.

Heneghan, J. B., Floyd, C. E. and Cohn, I. (1966). *J. surg. Res.* **6**, 24.

Holt, P. R., Haessler, H. A. and Isselbacher, K. J. (1963). *J. clin. Invest.* **42**, 777.

Horton, R. E. and Hickey, J. L. S. (1961). *Proc. Anim. Care Panel* **11**, 93.

Kellogg, T. F. (1966). *Symposium on Gnotobiotic Research*, University of Notre Dame, Notre Dame, Indiana, U.S.A.

Küster, E. (1915). *Arb. K. GesundhAmt.* **48**, 412.

Landy, J. J. and Sandberg, R. L. (1961). *Fedn Proc. Fedn Am. Socs exp. Biol.* **20**, 369.

Levenson, S. M. and Tennant, B. (1963). *Fedn Proc. Fedn Am. Socs exp. Biol.* **22**, 109.

Luckey, T. D. (1963). "Germfree Life and Gnotobiology", pp. 281, 494, 491, Academic Press, New York.

E

Luckey, T. D., Wagner, M., Gordon, H. A. and Reyniers, J. A. (1960). *Lobund Rep.* **3**, 176.

Luckey, T. D., Wagner, M., Reyniers, J. A. and Foster, F. L. (1955). *Fd Res.* **20**, 180.

Meyer, R. C., Bohl, E. H., Henthorne, R. D., Tharp, V. L. and Baldwin, D. E. (1963). *Lab. Anim. Care* **13**, 655.

Miyakawa, M. (1955). *Jap. J. med. Prog.* **42**, 553.

Mori, T. and Kato, S. (1959). *Tohoku J. exp. Med.* **69**, 197.

Newton, W. L. (1965). "Methods of Animal Experimentation", (W. I. Gary, ed.) Vol. 1, p. 215, Academic Press, N.Y.

Newton, W. L. and De Witt, W. B. (1961). *J. Nutr.* **75**, 145.

Nuttal, G. H. F. and Thierfelder, H. (1897). *Hoppe Seyler's Z. physiol. Chem.* **23**, 231.

Olson, G. B. and Wostmann, B. S. (1964). *Proc. Soc. exp. Biol. Med.* **116**, 914.

Phillips, B. P., Wolfe, P. A. and Gordon, H. A. (1959). *Ann. N.Y. Acad. Sci.* **78**, 183.

Pleasants, J. R. (1959). *Ann. N.Y. Acad. Sci.* **78**, 116.

Pleasants, J. R. (1966). *Ph.D. Thesis*, University of Notre Dame, Notre Dame, Indiana, U.S.A.

Pleasants, J. R. and Wostmann, B. S. (1962). *Proc. Indiana Acad. Sci.* **72**, 87.

Pleasants, J. R., Reddy, B. S. and Wostmann, B. S. (1966). *Symposium on Gnotobiotic Research*, University of Notre Dame, Notre Dame, Indiana, U.S.A.

Pleasants, J. R., Wostmann, B. S. and Zimmerman, D. R. (1964). *Lab. Anim. Care* **14**, 37.

Pleasants, J. R., Reddy, B. S., Zimmerman, D. R., Bruckner-Kardoss, E. and Wostmann, B. S. (1967). *Z. Versuchstierk.* **9**, 195.

Porter, G. and Lane-Petter, W. (1965). *Br. J. Nutr.* **19**, 295.

Reddy, B. S. and Wostmann, B. S. (1966). *Archs Biochem. Biophys.* **113**, 609.

Reddy, B. S., Wostmann, B. S. and Pleasants, J. R. (1967). *Fedn Proc. Fedn Am. Socs. exp. Biol.* **26**, 325.

Reddy, B. S., Pleasants, J. R., Zimmerman, D. R. and Wostmann, B. S. (1965). *J. Nutr.* **87**, 189.

Rice, E. E. and Beuk, J. F. (1953). *Adv. Food Res.* **4**, 233.

Rohovsky, M. W., Griesemer, R. A. and Wolfe, L. G. (1966). *Lab. Anim. Care* **16**, 52.

Reyniers, J. A. and Sacksteder, M. R. (1958). *Proc. Anim. Care Panel* **8**, 41.

Reyniers, J. A., Sacksteder, M. R. and Ashburn, L. L. (1964). *J. natn. Cancer Inst.* **32**, 1045.

Reyniers, J. A., Trexler, P. C. and Ervin, R. F. (1946). *Lobund Rep.* **1**, 1.

Smith, C. K. (1966). *Ph.D. Thesis*, University of Notre Dame, Notre Dame, Indiana, U.S.A.

Smith, C. K. and Trexler, P. C. (1960). *Int. Congr. Nutr. V*, Washington, D.C., p. 26.

Van Durren, B. L., Nelson, N., Orris, L., Palmes, E. D. and Schmidt, F. L. (1963). *J. natn. Cancer Inst.* **31**, 41.

Waxler, G. L., Schmidt, D. A. and Whitehair, C. K. (1966). *Am. J. vet. Res.* **27**, 300.

Whitehair, C. K. and Waxler, G. L. (1963). *Lab. Anim. Care* **13**, 665.

Windmueller, H. G., Ackerman, C. J. and Engel, R. W. (1956). *J. Nutr.* **60**, 527.

Wolfe, L., Griesemer, R. and Rohovsky, M. (1966). *Lab. Anim. Care* **16**, 364.

Wostmann, B. S. (1959). *Ann. N.Y. Acad. Sci.* **78**, 175.

Wostmann, B. S. (1961). *Ann. N.Y. Acad. Sci.* **94**, 272.

Wostmann, B. S. and Bruckner-Kardoss, E. (1966). *Proc. Soc. exp. Biol. Med.* **121**, 1111.

Wostmann, B. S. and Kellogg, T. F. (1967). *Lab. Anim. Care.* **17**, 589.

Wostmann, B. S. and Pleasants, J. R. (1959). *Proc. Anim. Care Panel* **9**, 47.

Zimmerman, D. R. and Wostmann, B. S. (1963). *J. Nutr.* **79**, 318.

Chapter 5

Characteristics of the Germ-free Animal

JULIAN R. PLEASANTS

*Lobund Laboratory, Department of Microbiology,
University of Notre Dame, Notre Dame, Indiana, U.S.A.*

The characteristics of the germ-free animal include those which characterize the animal as animal and those which characterize it as germ-free. It is natural that scientific interest should first centre on those aspects of the germ-free animal which are peculiar to the germ-free state. It is these aspects which provide insight into the effects of host-microbe interactions. Such insights are possible and meaningful, however, only if the germ-free animal is a successful animal, able to carry out the essential functions of life. Otherwise the differences we find between germ-free and conventional animals may be due to a failure of animal function rather than to the absence of microbial stimulation. The study of the germ-free animal as animal is basic to its study as germ-free.

Furthermore, the study of the germ-free animal is the study of the animal as such. It offers for the first time a demonstration of what the pure animal can do. It constitutes an asymptote for all the other ecological systems in which microbes play a role and thus provides fundamental data for anatomy, physiology, nutrition and biochemistry. From this viewpoint, the similarities between germ-free and conventional animals are at least as important as the differences. Once established, the similarities then simplify investigations of the effects of the microflora, e.g. on host defence mechanisms, for in such cases we would like to consider the host's general functioning as a constant, with host-microbe interaction as the variable.

It is the object of this chapter to show that the germ-free animal can be characterized as a successful animal, able to meet the demands of animal life. It is the function of other sections in this chapter to deal with those differences which characterize the animal as germ-free. Some of them are not only expected but desired. The differences in defence apparatus between the two groups are the very source of their value in immunological research. When more generalized functions differ, e.g. blood circulation, basal metabolism and intestinal wall development, this is not entirely unexpected, since it may reflect the magnitude of the animal's "defence budget". It would be dangerous to conclude, however, without direct evidence, that the differences are due only to a reduction in defence expenditure. Other possible explanations must and will be considered later (Chapter 6) along with a discussion of an unexpected and undesired peculiarity of germ-free rodents and rabbits, an extreme distention of the caecum.

Any markedly different generalized function could affect the response of germ-free animals to special experimental conditions. Under normal rearing conditions, however, the germ-free animal has proved itself a successful animal, capable of meeting the demands of general homeostasis, growth, reproduction, survival and the usual non-microbial stresses and

TABLE I

Growth of germ-free animals

Germ-free growth reported as equal or superior to conventional growth
  Chicken (cf. Chapter 3)
  Monkey (Reyniers and Trexler, 1943; Wolfe et al., 1966)
  Japanese quail (Reyniers and Sacksteder, 1960)
  Pig (Landy et al., 1961)

Germ-free growth usually reported as very close to conventional rate
  Rat (Gustafsson, 1959; Gordon, 1959)
  Mouse (Newton, 1965)
  Dog (Griesemer and Gibson, 1963; Heneghan et al., 1966)
  Cat (Rohovsky et al., 1966)
  Turkey (Forbes et al., 1958; Luckey et al., 1960)
  Calf (Griesemer, 1966)
  Burro (Griesemer, 1966)

Germ-free growth consistently reported lower than best conventional
  Guinea pig (Miyakawa et al., 1958; Tanami, 1959; Phillips et al., 1959; Horton and Hickey, 1961; Newton and De Witt, 1961; Abrams and Bishop, 1961; Pleasants et al., 1963)
  Rabbit (Pleasants et al., 1963; Wostmann and Pleasants, 1959; Reddy et al., 1965)
  Kid (Kuster, 1915; Smith, 1966)
  Lamb (Smith, 1966)

---

strains. It is these basic similarities of the germ-free and conventional animal which will be summarized below.

I. Growth

Before the advent of the germ-free animal, the growth potential of a species could be estimated only from the performance of the best conventional animal. The first studies on antibiotics and growth had already indicated, however, that the growth rate of conventional animals was not a stable point of reference. Effects of antibiotics on growth could be demonstrated in some laboratories but not in others and the reason seemed to lie in the variability of the conventional animal rather than in that of the antibiotic-fed animal. The same problem arises when we try to assess the growth rate of germ-free animals. We must use the conventional animal for a point of reference because it is the only one in our experience, but comparisons of germ-free and conventional growth may differ from one laboratory to the next and the differences may lie primarily in the instability of the conventional referent. Thus the data in Table I must be interpreted with this reservation in mind.

Chickens have been most frequently observed to show superior growth under germ-free conditions (see Chapter 5 for more details of chicken growth). The growth of recent germ-free rats at Lobund Laboratory on diet L-462 (Wostmann, 1959) is shown in Fig. 1. The growth is much superior to that reported earlier (Gordon, 1959a). Correction of over-crowded conditions appears to have been responsible for a simultaneous increase in growth rate, decrease in adrenal size and increase in thymus size, changes which suggest that the previously reported characteristics were due to stress (Gordon, 1959a; Gordon and Wostmann, 1960). The germfree rats still grow at a slightly lower rate than the best conventional animals, although there have been periods when the conventional rate was lower than the germ-free rate because it was also lower than the best conventional rate. The growth index for germ-free rats given in Table II shows that it equalled or exceeded the acceptable standard given by the National Research Council (Kellogg, 1966). This recent growth was obtained on diet L-484 (see Chapter 4).

Figure 2 shows the slower growth of germ-free guinea pigs. A rate 20–25% below the best conventional rates has been observed on a wide variety of diets (Miyawaka et al., 1958; Tanami, 1959; Phillips et al., 1959; Horton and Hickey, 1961; Newton and De Witt, 1961; Abrams and Bishop, 1961) and among both surgically derived and mother-suckled animals. There have been individual germ-free guinea pigs and rabbits, however, which gained at conventional rates, suggesting that diets may be just marginal in adequacy or digestibility or that overcrowding may be contributing to a reduction in average gain as it did for the first germ-free rats (Gordon, 1959a). It should be noted that the digestive tracts of

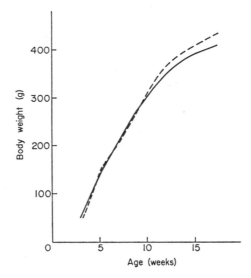

FIG. 1 Growth of male Lobund Wistar rats (unpublished data) fed autoclaved diet L-462 (Wostmann, 1959). Germ-free, ———; conventional, - - - - .

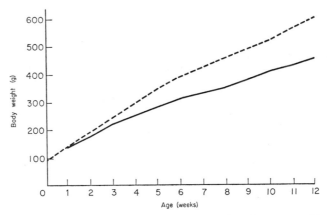

FIG. 2 Average body weight (in g) of guinea pigs (sexes combined) given autoclaved diet L-477 (Pleasants *et al.*, 1963) during the first 12 weeks of life. Germ-free, ———; conventional, - - - - .

guinea pigs and rabbits are much more specialized to permit prolonged microbial action on the digesta than is true of the other laboratory species. Thus none of the diets so far fed may have been sufficiently modified to compensate for this lack of activity. While ruminants, of course, have even greater anatomical specialization to promote microbial action on the digesta, it should be noted that sheep, goats and cattle have been maintained germ-free only during the pre-ruminant period of their lives when they normally live on milk, rather than on microbially-altered feedstuffs.

In summary, growth of germ-free animals is adequate in most species, well within conventional range, and tending to the upper extreme of the range in some species. Even guinea pigs and rabbits, which grow more slowly when germ-free, attain normal reproductive maturity.

TABLE II

Performance of germ-free rats on natural-type diet L-484, compared with that on other diets and with NRC standards

	Previous Lobund data	L-484	NRC standards
Fertility[a]	80%	87%[b]	90%
Pups/litter	8·0	8·9[c]	8·5
Pups weaned (No.)	6·4	6·2	7·6
Pups weaned (%)	80	70	90
Growth constant[d]		♂3·60	3·47
		♀2·80	2·77

[a] Percentage of females becoming pregnant during 15 day exposure to male.
[b] 156 matings.
[c] 120 litters; productive span was 5 to 6 litters.
[d] $K = \dfrac{\log W_2 - \log W_1}{1/T_1 - 1/T_2}$ (Zucker et al., 1941).

II. Reproduction

Germ-free rats are now in their 22nd generation at Lobund Laboratory, and germ-free mice are in their 28th generation. This record is made all the more remarkable by the fact that each of the oldest lines began from a single litter taken from a randomly-bred colony; the early generations had neither the advantages of random breeding nor the advantages of inbreeding. Table II shows the fecundity of germ-free rats on diet L-484 (for composition, see Chapter 4). Table III shows the reproduction records of germ-free mice and guinea pigs. In each case, reproduction approached that of conventional animals. Rabbits maintained into the third generation showed a progressive and finally total loss of fertility, but a harmful effect of inbreeding on rabbit fertility is well-known and we therefore do not know if germ-free status played any part in this result. Although germ-free chickens have reproduced (Reyniers et al., 1949), no attempt was made to continue the reproduction. Japanese quail have reproduced into the third generation (Reyniers and Sacksteder, 1960) but the statistics of their reproduction have not yet been given. Reproduction by the larger germ-free species has not been reported, since they have not been maintained germ-free long enough for reproduction.

Some strains of germ-free rats and mice have shown impaired fertility on particular diets that were adequate for conventional controls. The same was true of the first germ-free rabbits (Wostmann and Pleasants, 1959) and of hand-fed rats and mice. Those who have worked with a variety of

TABLE III

Comparative reproduction of germ-free and conventional mice[a] and guinea pigs[b]

Species	Status	Number of females	Months of observation	Number of litters	Number of weaned young	Weaned young per female per month
Mouse	Germ-free	9	6	36	191	3·5
Mouse	Conventional	9	6	37	203	3·8
Guinea pig	Germ-free	4	5[c]	8	14	0·70
Guinea pig	Conventional	12	4[c]	18	39	0·83

[a] Newton (1965).
[b] Pleasants et al. (1963).
[c] Average period of observation.

rodent species and strains are inclined to correlate lowered fertility with extreme caecal distention, whether caused by hand feeding, a particular solid diet, or a special strain susceptibility to caecal enlargement. The first reproduction in germ-free rabbits was obtained only after caecal ligation to increase both survival and breeding (Pleasants et al., 1963). In later experiments a solid diet which caused less extreme caecal distention than earlier diets and provided better haemoglobin production (Reddy et al., 1965) permitted reproduction without the necessity of caecal ligation.

Under germ-free conditions it is sometimes difficult to meet the known standards of housing, lighting and genetic management for maximum reproduction and it is not clear how well these standards have been met in cases of lower fertility in germ-free animals. The delay in breeding seen in surgically-derived germ-free guinea pigs was not observed in later generations (Pleasants *et al.*, 1963); thus it was not a characteristic of the germ-free state but probably resulted from the fact that guinea pigs reared in a breeding colony learn efficient mating procedures earlier than those reared in isolation (Young *et al.*, 1964).

III. Survival

Germ-free animals may be divided into three groups in terms of their ability to survive.

A. RATS AND MICE

These species outlive their conventional counterparts (Newton, 1965; Gordon, 1959b; Gordon *et al.*, 1966). This is true even though most of the causes of death in the germ-free animal result from conditions peculiar to the germ-free state, such as the distended caecum (cf. Chapter 6). The hazards of microbial association outweigh the hazards of not being associated with microbes.

B. GUINEA PIGS AND RABBITS

These species have not been maintained long enough to permit comparison of life spans, but limited data are available about early mortality (Pleasants, unpublished). Among thirty-four guinea pigs born alive to germ-free mothers and normally suckled by them, there was no perinatal mortality such as sometimes occurs in conventional colonies. At 2 months of age the first deaths occurred, two animals dying from anal prolapse, a condition to which germ-free guinea pigs housed in groups seem especially susceptible. Most of the remaining animals were used for experiment or were accidentally contaminated, but there were three spontaneous deaths from volvulus of the enlarged caecum, at ages of 8, 13, and 17 months. Among surgically derived guinea pigs fed diet L-447 autoclaved with 25% added moisture, caecal distention developed some months later than in the naturally born, normally suckled animals and one surgically derived animal lived to 28 months. In both groups, however, there was a progressive increase with age in the proportionate weight of the caecum and eventual death from volvulus. Extrapolating from these data, one might say that in the very early life of the guinea pig, the hazards of conventional life outweigh the hazards of germ-free life. During some later period, the hazards of germ-free life outweigh the conventional. Since, however, the guinea pig is the one rodent species so far studied in which caecal distention can be prevented, at least temporarily, by dietary means (Horton and

Hickey, 1961; Pleasants *et al.*, 1963; Pleasants *et al.*, 1967) it seems probable that further research will eliminate this particular hazard for the guinea pig.

Among twenty-five naturally-born, normally-suckled germ-free rabbits, two were lost at weaning, one with a ruptured intestine, one with no obvious cause of death. Among nineteen animals kept germ-free past maturity, three died from volvulus of the distended caecum, at ages of 11, 18 and 22 months. These were the only spontaneous deaths that could be attributed to a condition peculiar to the germ-free state. The incidence of volvulus in both naturally-born and surgically-derived germ-free rabbits was much reduced after adoption of solid diet L-478 (Reddy *et al.*, 1965) on which the caecal volume is less than half that observed on earlier diets. One surgically derived rabbit lived for 35 months on diet L-478.

Thus, in rabbits as in guinea pigs, there is hope that dietary modifications can eliminate the disabilities which now make germ-free life more hazardous than conventional life after maturity. The fact that caecal distention causes greater mortality in germ-free guinea pigs and rabbits than in germ-free rats and mice may be due to the fact that the caecum of guinea pigs and rabbits is a large organ even in the conventional state. If germ-free status multiplies the caecal weight by 5 (a rough approximation), this brings the caecal size of germ-free rats and mice to 5% of body weight but brings the caecal size of germ-free guinea pigs and rabbits to 25% of body weight. Germ-free ruminants also required special diets, since mineral metabolism was particularly affected by germ-free status. For further discussion of altered mineral metabolism in the germ-free state see Chapters 4 and 8. A variety of kidney lesions, especially suggestive of calcification, have been observed in germ-free rodents and rabbits (Wostmann and Pleasants, 1959) and lambs and kids (Smith, 1966) but their incidence and severity have progressively decreased as diets have been improved. Among lambs and kids maintained germ-free up to 4 months of age, there was some evidence in the lungs of an allergic response, possibly to the cow's milk on which they were fed. This was more severe in the germ-free animals. An allergic response in germ-free rabbits hand-fed a diet based on cow's milk has also been reported (Coates and O'Donoghue, 1967). While unexpected, these reactions are not surprising. It has never been considered desirable to use the hand-fed generations of germ-free mammals for general experimentation.

Dogs (Heneghan *et al.*, 1966), cats (Rohovsky *et al.*, 1966) and monkeys (Wolfe *et al.*, 1966) have been maintained germ-free for at least 10 months. No special hazards of germ-free life (other than altered dietary requirements) had appeared in that time. Thus caecal distention is not a characteristic of germ-free animals in general but only of four species, rat, mouse, guinea pig and rabbit. These four species happen to be those in which coprophagy is a way of life.

If special germ-free hazards do not exist in a species, as seems likely for most of them, or if the hazards can be eliminated where they presently exist (rodents and rabbits) then every germ-free species will outlive its conventional counterparts, since the germ-free animals will be exposed to fewer hazards.

IV. Morphology and Physiology

The germ-free animal is clearly able to meet its own needs of growth, reproduction and survival if provided with a suitable diet and environment. From the normality of these functions, however, we cannot assume *a priori* that the germ-free animal is just like the conventional animal in all aspects except those specifically involved in anti-microbial defence. Active defence against microbial action involves tissues, organs and systems which also serve other functions, e.g. the digestive system, heart, liver and adrenals. If these do not have to carry out anti-microbial activity, they may be altered in form and functional capacity in ways that could affect their other functions and their reponses to experimental stress. Such differences must be investigated not only to discover how much of an animal's structure and function is affected by bacteria but also to determine if the germ-free animal is suffering from any borderline deficiency which would make it more susceptible to special stress. In any experiment which involves measuring anatomical and physiological parameters of the germ-free animal in the presence of experimental variables, we need to know if those parameters have already been influenced by the germ-free state itself.

Fortunately for the present-day experimenter, an extensive survey of many morphological and biochemical characteristics of the germ-free animal has already been carried out. Table IV gives an idea of the range of these surveys. Much of the original data may be found in papers by Gordon (1959a); Gordon and Wostmann (1960); Pleasants *et al.* (1967); Reyniers *et al.* (1960) and Luckey (1963) as well as in the other chapters of this book.

The most important conclusion to be drawn from such surveys is that the organs and tissues not usually in contact with bacteria or indirectly contributing to anti-microbial defence, all those parts of the body which Gordon (1959a) has called the organs of the internal environment, are essentially similar in germ-free and conventional animals. These facts, as well as the ability of the germ-free animal to carry out the normal functions of life, prove that the common laboratory species have not evolved any genuinely symbiotic relationship with microbes and that they can maintain homeostasis over a broad range of external milieux.

However, three kinds of buffer stand between the constancy of the internal milieu and the inconstancy of the external milieux. In the first line stand the gastrointestinal tract (and to a much lesser extent the skin and respiratory tract) exposed to an environment which is teeming with

TABLE IV

Comparative surveys of germ-free and conventional animals
(species best studied: chicken, rat, guinea pig, mouse)

Organs studied			
Whole body	Oesophagus, crop	Liver, bile	Blood
Skin	Proventriculus	Pancreas	Thymus
Fur, feathers	Gizzard	Kidney, urine	Spleen
Skeleton	Stomach	Brain	Lymph nodes
Musculature	Small intestine	Pituitary	Trident
Fat depots	Caecum	Thyroid	Bursa
Lungs, trachea	Large intestine	Gonads	
Heart	Cloaca, rectum	Adrenals	
Aspects studied			
Appearance	pH	Proteases	Haemoglobin, haematocrit
Wet weight	Redox	Amylases	Iron-binding capacity
Dry weight	B vitamins	Lipases	Plasma Fe, Cu, Mn
Ash, P, Na, K	vitamin C	Disaccharidases	RBC, WBC, differential
Lipids	Carnitine	Monoamine oxidase	Clotting time
Cholesterol	Histamine	DPNH diaphorase	Clotting factors
Nitrogen	Serotonin	Phosphatases	

microbes and has been altered in innumerable ways by those microbes. In the second line stands the whole specialized defence apparatus of the host. In the third line stand those organs, such as liver, heart and adrenal glands, whose total burden is altered by the microbial content of the external milieu. The germ-free animal is bound to differ from the conventional animal in aspects of these three buffer systems and it does. If it did not, microbes would be an irrelevant part of the environment.

The similarities and differences in those systems must be experimentally evaluated. The smaller liver, which has been observed, though not invariably, in germ-free rats, mice and chickens (Gordon, 1959a) could obviously affect the capacity of the germ-free animal to metabolize certain substances. The smaller heart seen in germ-free rats and mice (Gordon, 1959a) could also play a role. Miyakawa (1966) found smaller adrenal glands in germ-free rats, the zona glomerulosa and zona fasciculata being less well developed under germ-free conditions. Studies of basal metabolism and blood circulation in germ-free animals show reductions in these measurements of function (Table V). While it is tempting to assume that these anatomical and functional changes in the germ-free animal indicate the magnitude of the "defence budget" in conventional animals, other possible explanations will be discussed in Chapter 6.

Other practical differences between germ-free and conventional animals which can affect experimental results include: rate of absorption of ingested materials (Heneghan, 1963; Laroche *et al.*, 1964), response to anaesthesia (Quevauviller *et al.*, 1964) and voluntary intake of food (Luckey *et al.*, 1954). In the last mentioned instance, a diet designed to produce liver necrosis failed to do so in germ-free rats because they ate twice as much as the conventional rats. The germ-free animals developed symptoms only when their intake was restricted to that of the conventional animals. Such possibilities must be considered in the design of experiments involving germ-free animals.

TABLE V

Metabolic activity and blood circulation in germ-free rats.
Reduction of germ-free level below conventional level

	Reduction expressed as % of conventional level	Reference
Basal metabolic rate	20	Desplaces *et al.* (1963)
	15–20	Levenson *et al.* (1966)
Iodine uptake by thyroid (normal diet)	50	Levenson *et al.* (1966)
Cardiac output (ml/min/kg B.W.)	32	Gordon *et al.* (1963)
Blood volume (ml/100 gm B.W.)	22	Gordon *et al.* (1963)
Arterial blood flow to liver	50	Gordon *et al.* (1963)

The phenomenon of caecal distention which occurs in a few germ-free species is so abnormal that it does at times overburden the buffer systems and threaten homeostasis itself. This raises the question of whether certain systemic characteristics of the germ-free rodent are actually secondary responses to caecal enlargement, rather than reflections of a reduced demand for anti-microbial defence. This question will be dealt with in a later chapter. It should be noted here that caecal distention occurs in only four out of some sixteen vertebrate species reared germ-free. Unfortunately these happen to be the four most important species of laboratory mammals. They are also just those species which have specialized both caecum and behaviour to promote and benefit from microbial action on caecal residues. Our ability to compensate for other microbial contributions to the host, such as vitamin production, promotion of iron absorption in the rabbit (Reddy *et al.*, 1965) and hydrolysis of lactose in the piglet (Landy et al., 1961) and our ability to influence caecal size to a considerable extent by diet, give us grounds for believing that there are no abnormalities indissolubly linked to the germ-free state. There are only abnormalities linked to the state of our ignorance about the role of bacteria in animal life.

The germ-free animal will provide the answer to its own problems. In fact, as we come to understand the exact biochemical basis of the differences between germ-free and conventional animals, it should become possible to eliminate or exaggerate those differences at will, using pure chemical compounds for this purpose, rather than introducing the biochemical complexity which characterizes even a single living microbial associate. Thus the experimental animal could be tailor-made for a particular type of experiment.

Meanwhile, the basic success of the germ-free animal as a functioning animal means that it holds the answers to many other problems not its own. Both the similarities of its internal milieu and the differences of its buffering systems are keys to the unlocking of biological and microbiological secrets.

REFERENCES

Abrams, G. D. and Bishop, J. E. (1961). Univ. Mich. med. Bull. 27, 136–147.

Coates, M. E. and O'Donoghue, P. N. (1967). Nature, Lond. 213, 307–308.

Desplaces, A., Zagury, D. and Sacquet, E. (1963). C.r. hebd. Séanc. Acad. Sci., Paris 257, 756–758.

Forbes, M., Supplee, W. C. and Combs, G. F. (1958). Proc. Soc. exp. Biol. Med. 99, 110–113.

Gordon, H. A. (1959a). Ann. N.Y. Acad. Sci. 78, 208–220.

Gordon, H. A. (1959b). Gerontologia 3, 104–114.

Gordon, H. A. and Wostmann, B. S. (1960). Anat. Rec. 137, 65–70.

Gordon, H. A., Wostmann, B. S. and Bruckner-Kardoss, E. (1963). Proc. Soc. exp. Biol. Med. 114, 301–304.

Gordon, H. A., Bruckner-Kardoss, E. and Wostmann, B. S. (1966). J. Geront. 21, 308–387.

Griesemer, R. A. (1966). Int. Congr. Microbiol. IX Symposia 287–289.

Griesemer, R. A. and Gibson, J. P. (1963). Lab. Anim. Care 13, 643–649.

Gustafsson, B. E. (1959). Ann. N.Y. Acad. Sci. 78, 17–28.

Heneghan, J. B. (1963). Am. J. Physiol. 205, 417–420.

Heneghan, J. B., Floyd, C. E. and Cohn, I. (1966). J. surg. Res. 6, 24–31.

Horton, R. E and Hickey, J. L. S. (1961). Proc. Anim. Care Panel 11, 93–106.

Kellogg, T. F. (1966). Symposium on Gnotobiotic Research, p. 6 (abstract), University of Notre Dame, Notre Dame, Indiana, U.S.A.

Küster, E. (1915). Arb. K. GesundhAmt. 48, 1–79.

Landy, J. J., Growden, J. H. and Sandberg, R. L. (1961). J. Am. med. Ass. 178, 1084–1087.

Laroche, M.-J., Cottart, A., Sacquet, E. and Charlier, H. (1964). Annls pharm. fra. 22, 333–338.

Levenson, S. N., Kan, D., Lev, M. and Doft, F. S. (1966). *Fedn Proc. Fedn Am. Socs exp. Biol.* **25**, 482 (abstract).

Luckey, T. D. (1963). "Germ-free Life and Gnotobiology", Academic Press, New York.

Luckey, T. D., Reyniers, J. A., Gyorgy, P. and Forbes, M. (1954). *Ann. N.Y. Acad. Sci.* **57**, 932–935.

Luckey, T. D., Wagner, M., Gordon, H. A. and Reyniers, J. A. (1960). *Lobund Rep.* **3**, 176–182.

Miyakawa, M. (1966). *Internat. Congr. Microbiol.* IX Symposia 291–298. Ivanovski Inst. Virol., Moscow, Russia.

Miyakawa, M., Iijima, S., Kishimoto, H., Kobayashi, R., Tojima, M., Isomura, N., Asano, M. and Hong, S. C. (1958). *Acta path. jap.* **8**, 55–78.

Newton, W. L. (1965). *In* "Methods of Animal Experimentation". (W. I. Gay, ed.), Vol. 1, pp. 215–271, Academic Press, New York.

Newton, W. L. and Dewitt, W. B. (1961). *J. Nutr.* **75**, 145–151.

Phillips, B. P., Wolfe, P. A. and Gordon, H. A. (1959). *Ann. N.Y. Acad. Sci.* **78**, 116–126.

Pleasants, J. R., Zimmerman, D. R., Reddy, B. S. and Wostmann, B. S. (1963). *Proc. Gnotobiote Workshop and Symposium*, pp. 36–50B. Ohio State University, Columbus, Ohio, July, 1963.

Pleasants, J. R., Reddy, B. S., Zimmerman, D. R., Bruckner-Kardoss, E. and Wostmann, B. S. (1967). *Z. Versuchstierk.* **9**, 195–204.

Quevauviller, A., Laroche, M. J., Cottart, A., Sacquet, E. and Charlier, H. (1964). *Annls pharm. fr.* **22**, 339–344.

Reddy, B. S., Pleasants, J. R., Zimmerman, D. R. and Wostmann, B. S. (1965). *J. Nutr.* **87**, 189–196.

Reyniers, J. A. and Trexler, P. C. (1943). *In* "Micrurgical and Germ-free Technique" (J. A. Reyniers, ed.). pp. 114–143, Charles C. Thomas Co., Springfield, Illinois, U.S.A.

Reyniers, J. A. and Sacksteder, M. R. (1960). *J. natn Cancer Inst.* **24**, 1405–1421.

Reyniers, J. A., Trexler, P. C., Ervin, R. F., Wagner, M., Luckey, T. D. and Gordon, H. A. (1949). *Lobund Rep.* **2**, 119–148.

Reyniers, J. A., Wagner, M., Luckey, T. D. and Gordon, H. A. (1960). *Lobund Rep.* **3**, 7–171.

Rohovsky, M. W., Griesemer, R. A. and Wolfe, L. G. (1966). *Lab. Anim. Care* **16**, 52–59.

Smith, C. K. (1966). Ph.D. Thesis, University of Notre Dame, Notre Dame, Indiana, U.S.A.

Tanami, J. (1959). *J. Chiba Med. Soc.* **35**, 1–21.

Wolfe, L., Griesemer, R. and Rohovsky, M. (1966). *Lab. Anim. Care* **16**, 364–368.

Wostmann, B. S. (1959). *Ann. N.Y. Acad. Sci.* **78**, 175–182.

Wostmann, B. S. and Pleasants, J. R. (1959). *Proc. Anim. Care Panel* **9**, 47–54.

Young, W. C., Goy, R. W. and Phoenix, C. H. (1964). *Science, N.Y.* **143**, 212–218.

Zucker, T. F., Hall, L., Young, M. and Zucker, L. (1941). *J. Nutr.* **22**, 123–138.

Chapter 6

Is the Germ-free Animal Normal?

A Review of its Anomalies in Young and Old Age

H. A. GORDON

Department of Pharmacology, College of Medicine,
University of Kentucky, Lexington, Kentucky, U.S.A.

I. Introduction

Since the beginning of experimental work with germ-free animals, certain characteristics and distinct types of response became apparent in this form of life. The lack of development of some elements of the cellular and humoral defensive system seemed to be a plausible consequence of the greatly reduced exposure to antigens. By the same token, the germ-free animal's increased resistance to some stressor agents (e.g. X-irradiation), or its longer life span have been attributed to the relief from the conventional microbial burden. In these instances the contribution of the microbial flora appeared as a complicating, or as an outright detrimental factor in the conventional host's life. Yet, we argued, the absence or presence of the flora could not be too far-reaching in its overall effects on the host in view of the normal growth, development and reproduction which were observed in germ-free animals when compared to their conventional controls. This reassurance came particularly after conditions of the germ-free experiment could be standardized and a plentiful supply of young and young adult, normally-born and mother-suckled germ-free animals became available for investigation. In addition, many morphological and functional details, originating mainly from areas of the body that are beyond the immediate microbial exposure, proved similarity between germ-free and conventional animals. Finally there were observations, some of them quite old, which suggested that there is also something basically different, or perhaps even wrong, with germ-free animals. Although ultimately caused by the lack of bacteria, these differences could not be readily explained in terms of the lack of microbial stimulation and a resultant tissue under-development as was the case in the defensive system. This group of observations, which indicated more or less pronounced departures from conventional standards in structure, chemical composition and function of various organs in germ-free animals, has lately expanded. Some of these differences, like the ones resulting from the sparing effect of germ-free life on various intestinal degradation processes, did not appear to influence particularly the physiological economy of the host. Other changes, like the enlarged caecum and its consequences seemed patently detrimental to the animal's health.

In this context, a comment made by Sacquet (1959) is particularly meaningful: "Thus, the question previously posed by Pasteur (1885),—is the aseptic animal possible?—must be substituted by—is the aseptic animal normal?—or, on the contrary,—does it suffer from some insufficiency that we should know of, before we use it as an experimental subject?".

The present article attempts to give an answer to these questions. It concentrates on the anomalies of germ-free life, i.e. on the aberrations from physiological normality that develop when the animal host is uninfluenced by microbial associates. It does not deal with nutritional and immunological issues, effects of surgical delivery, hand-feeding, confinement in isolators and other non-bacterial variables that may affect the germ-free animal (Pleasants, 1959). For this reason it concentrates only on animal species that are available today in form well adapted to germ-free life. This limits our discussion to rats and mice that have been normally born and mother-suckled in germ-free colonies over generations. Some of these aspects have been recently reviewed (Gordon, 1965a). In addition to published material, this article also mentions occasional observations which invariably accumulate in laboratories devoted to this field and which may be of some value to readers with practical interest.

II. The Gastrointestinal Tract

A. THE INTESTINAL WALL

To the naked eye, the gastrointestinal tract of germ-free animals appears somewhat thin-walled and well filled. Against this more or less conventional appearance of the gut, the distended and enlarged caecum has been the landmark of germ-free life. actually, this was one of the first observations made in germ-free animals (Nuttal and Thierfelder, 1896) and it still remains the main riddle of germ-free life. The colon is more or less normal in appearance, although the formation of faecal pellets is less clearly indicated than in conventional animals.

In freshly excised specimens, the wall of the small intestine shows reduced wet weight when expressed per unit body weight in comparison to conventional controls (even after correction for caecal contents). Dry weight of the wall tissue is usually higher than normal (Gordon et al., 1966b). A study of tissue distribution in histological specimens (Gordon and Bruckner-Kardoss, 1961a) indicates that the proportion of epithelium is usually normal or slightly elevated. The lamina propria is always considerably reduced in germ-free specimens. This may be one of the main causes of reduction of intestinal weight in these animals. In this tissue component considerable reduction of elements of the reticulo-endothelial system (RES) is also apparent. This may be the cause of the lower values of deoxyribonucleic acid which Combe et al. (1965) found in the intestinal wall of germ-free rats. The mucosal surface area of the small intestine in germ-free animals appears to be lower. In rats a reduction of approximately 30% was found which was most pronounced in the mid and lower parts of the small intestine. The reduction of the mucosal surface area, in comparison to conventional controls, was fairly well paralleled by the reduction in dry matter per unit length of the intestinal wall (Gordon and Bruckner-Kardoss, 1961b). Abrams et al. (1963) found significantly lower cell renewal rates in the epithelium, lamina propria and Peyer's patches of the intestine in germ-free animals. Phillips et al. (1961) reported that germ-free mice have higher levels of 5-hydroxytryptamine (5-HT) in the tissue of the small intestine than do conventional controls. Beaver and Wostmann (1962) observed that histamine in the same organ of germ-free rats and mice tends to be lower than in conventional controls. In the case of 5-HT, conditions were found generally reversed.

In contrast to the small intestine, the wet weight of the wall of the caecum, i.e. the caecal sac proper, is always elevated in germ-free conditions (Gordon et al., 1966b). Its percentage dry weight also appears higher. The caecal wall, due to the greatly increased contents, is extremely thin (Fig. 1). Tissue distribution measured in the caecal wall of young adult germ-free and conventional mice in our laboratory with the "random shot" method of Chalkley (1943) indicated that the greater wet weight of the germ-free caecal sac (1·9 : 1 compared with the conventional in this instance) was caused by a proportionate increase of the mucosa and

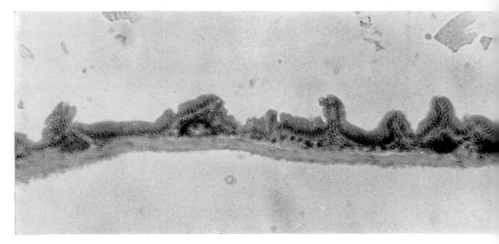

A

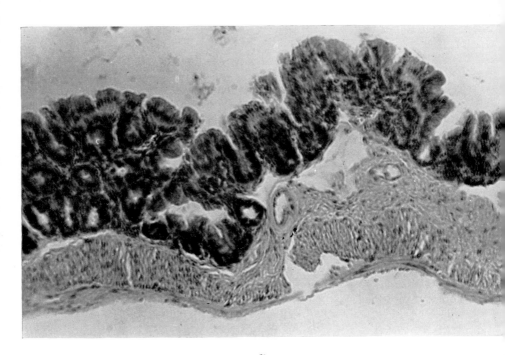

B

Fig. 1 Caecal wall (margo libera) of germ-free (A) and conventional (B) mice (males, 80 days old). Haematoxylin-eosin. Oc. 10 ×, Obj. 10 ×.

sub-mucosa (1·6:1) and by a considerable increment of muscular tissue (3·3:1). Subdividing the muscle into nucleo- and cytoplasmic components, it was indicated that in the germ-free state, the former shows only a slight increase over the conventional value (1·6:1) while the bulk of the increment rests with the cytoplasmic component (4·3:1). The individual smooth muscle cells appeared very elongated and hypertrophied in the germ-free specimens. Jervis and Biggers (1964) observed that in germ-free mice the caecal mucosa is generally thinner than in the corresponding regions of conventional mice, but it is raised in short irregular villi over its surface. They also found elevated levels of alkaline phosphatase and

TABLE I

Maximum relaxation (mm) of rat intestinal strips[a] in vitro[b]
caused by l-adrenaline[c]

	Germ-free			Conventional		
	N	M	S.D.	N	M	S.D.
Caecum	12	20	7	12	32	10
Colon	8	45	13	10	44	8

[a] Original length approx. 30 mm; lever magnification 1:4; counterweight 0·5 g.
[b] Organ bath: Vane, J. R. (1957).
[c] Concentration of adrenaline in organ bath 1:10[6].

N = number of animals; M = arithmetic mean; S.D. = standard deviation.

reduced levels of acid phosphatase (which may be associated with defensive functions) in the epithelial lining of the germ-free caecum. Dupont et al. (1965), investigating the intramural nervous system of the caecum in germ-free rats, found structural differences in the primary plexus of Auerbach together with increased size of myenteric neurones, some of them reaching "absolutely monstrous" size. In these cells, diphosphopyridine nucleotide diaphorase activity was found to be reduced. As this enzyme appears to be indicative of the levels of oxidative energy metabolism of a tissue, these results suggest that this segment of the autonomic nervous system in the germ-free rat is less active than in conventional controls. Th. E. Staley (personal communication) attempted to assay the tone of intestinal smooth muscle from germ-free rats in our laboratory by recording the maximum relaxation of intestinal strips in vitro that can be produced by the action of adrenaline (here "tone" was defined as the state of continuous contraction that exists between maximum contraction and maximum relaxation of the muscle). Some results of this work, shown in Table I, indicated that the tone of the germ-free caecal muscle strips was reduced to approximately two-thirds that of conventional controls. There was no difference in tone between the colon strips of germ-free rats and conventional controls. Strandberg *et al.* (1966) found that the concentration of catecholamines, acetylcholine and histamine were of the same order of magnitude in the ceacal wall of germ-free rats as in the

TABLE II

Daily water consumption of germ-free and conventional mice[a]

| Date | Water consumed by 3 mice (ml/day) | | | | | |
| | Germ-free | | | Conventional | | |
	Jar 1	Jar 2	Jar 3	Jar 4	Jar 5	Jar 6
July 16	12·0	10·4	12·0	9·6	6·4	7·2
17	16·8	24·0	13·6	7·2	9·6	8·0
18	11·2	13·6	11·2	11·2	7·2	5·6
19	9·6	11·2	10·4	6·4	8·8	5·6
20	11·2	12·8	11·2	8·8	8·0	5·6
21	6·4	3·2	6·4	4·0	8·0	7·2
22	12·8	14·0	12·0	9·6	13·6	7·6
23	11·2	15·6	10·4	6·4	6·8	6·8
24	12·0	13·6	11·6	7·2	8·4	5·6
25	10·8	14·0	9·6	6·4	7·2	9·6
26	10·8	13·3	10·4	6·4	6·2	7·0
27	10·8	13·3	10·4	6·4	6·2	7·0
28	12·0	10·8	12·0	8·0	9·2	7·2
29	13·6	13·2	12·4	9·6	9·6	6·8
30	14·4	11·2	10·4	6·8	4·8	10·0
31	9·6	13·6	10·0	6·4	6·4	6·4
M	11·6	13·0	10·9	7·5	7·9	7·1
S.D.	2·1	3·9	1·3	1·8	2·0	1·8
During 16 day period:						
food consumed, g	91·0	97·0	84·0	70·4	72·1	66·0
Feces voided, dry, g	37·9	35·6	30·6	27·6	26·5	19·3

[a] Swiss-Webster strain, 12–13 weeks old, 25–30 g females, fed on sterilized L-462 diet and tap water *ad lib*. All animals were housed in germ-free-type isolators and in gallon churn jars on wire mesh, 3 mice per jar. The isolators were held in the same room under standardized environmental conditions.

corresponding tissue of conventional controls (for histamine in the caecal wall, this represents a corroboration of the previous findings of Beaver and Wostmann, 1962). Germ-free caecal strips *in vitro* were generally less sensitive to L-adrenaline, acetylcholine, histamine and 5-HT. An old experience among workers in this field is that germ-free animals (rats particularly) tend to develop intestinal volvuli. These are seen almost exclusively at the ileo-caecal-colic junction and their development may take days or longer. Ultimately, they lead to intestinal occlusion and death. Such changes, along with the excessive size of the caeca (amounting to 25% or more of the body weight) are seen most frequently in young adult animals that were originally surgically derived and hand-fed artificial diets.

TABLE III

Wet weight[a] and pH[b] of intestinal contents in germ-free and conventional mice[c]

	Wet wt M	S.D.	pH
Stomach			
Germ-free	1·48	0·14	4·1–4·9
Conventional	1·55	0·16	4·5–4·9
Small intestine, upper 50%			
Germ-free	1·43	0·43	6·2–6·4
Conventional	0·64	0·17	6·1–6·2
Small intestine, lower 50%			
Germ-free	2·55	0·38	6·6–6·7
Conventional	0·91	0·30	6·4–6·7
Caecum			
Germ-free	12·77	1·45	7·0–7·1
Conventional	0·68	0·06	6·4–6·5
Colon			
Germ-free	1·01	0·24	6·8–6·9
Conventional	0·54	0·24	6·4–6·7

[a] In g/100 g body weight, corrected for caecal contents.
[b] pH minima–maxima, measured in freshly emptied contents with Beckman glass and calomel electrodes.
[c] Similar to those in Table II.

B. STATE OF REPLENISHMENT OF THE GUT

Schottelius (1902), reporting on his germ-free chickens, wrote: "An interesting phenomenon is manifested by the fact that chickens kept in sterile conditions are continuously hungry and eat, digest, and defaecate all the time. These little animals . . . surpass in appetite and discharge of excrement the normally nurtured control animals by many times". In a less drastic sense, this impression is shared by other workers and applies also to rodents. Combe *et al.* (1965) reporting on 60–70 day old germ-free rats found approximately 10% more ingesta (on a dry matter basis) in comparison to conventional controls. An illustration of similar results obtained in our laboratory in collaboration with T. Z. Csáky, with germ-free mice, is given in Table II. The same table indicates that these animals also tended to drink more water. A typical distribution of intestinal contents is given in Table III, with their percentage dry matter in Fig. 2. Generally one finds the stomach of germ-free animals well filled and its moisture content occasionally more elevated. In the small intestine, the

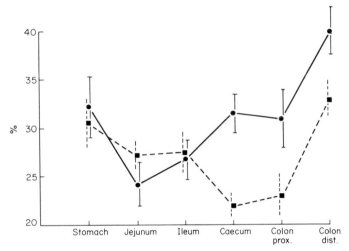

FIG. 2 Percentage dry matter of intestinal contents in various segments of the gastrointestinal tract of mice (same animals as given in Table II). Arithmetic means and standard deviations (cross bars). Germ-free, - - - -; conventional, ———.

quantity of contents is usually increased, extending over its entire course. The percentage dry matter of the intestinal contents, from the stomach to the ileocaecal valve, is essentially similar between germ-free and conventional animals. In the caecum the contents are increased 5–10 times, or even more, and considerably more liquid than in conventional controls, or for that matter, in normal animals in general. This may also be demonstrated by radiography following an enema of radiopaque material (Fig. 3). It must be emphasized that in the germ-free groups of some species (rodents in general) the caecal enlargement and the greater liquidity of the contents always occur concomitantly. In other species, (e.g. the chicken) the germ-free caecal contents are as liquid as in rodents

(Reyniers *et al.* 1960) but their volume is the same or only a little higher than in conventional controls. Another point of importance is in the mode of change in the percentage dry matter of intestinal contents between the

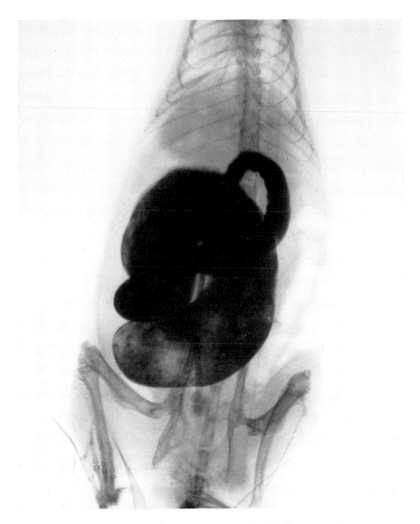

FIG. 3 Ex-germ-free adult male rat radiographed following administration of radiopaque enema within 30 min after exposure to conventional environment. Dorso-ventral exposure. Radiography: William Scruggs.

terminal junction of the ileum and the caecum. In conventional animals, as contents pass from the ileum to the caecum, there is a clearly discernible increase in percentage dry weight as a part of the general inspissation of intestinal contents from small intestine to rectum. In germ-free animals, passing from the terminal ileum to the caecum, there is an abrupt decrease in percentage dry matter. Thus, the contents change from a firm, pasty consistency in the ileum to semi-liquid in the caecum. Within the caecum of germ-free animals the most liquid contents may be found at or around the ileo-caecal-colic junction, while those at the apical portion of the caecum are somewhat less liquid. Once they reach the colon, a gradual thickening of the contents takes place. In the descending colon and rectum, the formation of faecal pellets is seen. Occasionally, the large bowel remains filled in its entire length with semi-liquid, unformed contents. The water absorption from the large intestine of germ-free animals persistently lags behind that of conventional animals, to the point that this state may be termed a mild diarrhoea. A tell-tale sign of this is the messy brown tail often seen in the germ-free rat or mouse. As mentioned in the previous paragraph, the nutritional history of the animals has an effect on the anomalies presently described. They are most pronounced in originally surgically delivered animals that have been fed on artificial formulae before weaning; caecal dilatation persists in the post-weaning period. Generally speaking, diets with higher roughage content usually result in larger caeca. Enlargement and the greater liquidity of caecal contents are clearly demonstrable in germ-free normally-born, mother-fed baby animals (Wostmann and Bruckner-Kardoss, 1959), as well as in similar but adult animals fed on liquid, amino acid type diets (Pleasants and Wostmann, 1962).

C. COMPOSITION[1] AND PROPERTIES OF CONTENTS

1. HYDROGEN ION CONCENTRATION, OXIDATION-REDUCTION POTENTIAL, OSMOTIC PRESSURE

The pH in the upper part of the intestinal canal is essentially similar in germ-free and conventional animals, but the pH in the lower bowel of germ-free animals is generally more alkaline than in conventional controls (Table III). According to Wostmann and Bruckner-Kardoss (1966), values in the caecum of germ-free rats average 1 pH unit higher. The oxidation-reduction potential of germ-free caecal contents was first measured in guinea pigs by Phillips and Wolfe (1959) and later in greater detail by Wostmann and Bruckner-Kardoss (1966). The germ-free values were found generally 250–300 mV more positive than in conventional controls. Observations on the osmolarity of intestinal contents in germ-free and conventional rats are given in Table IV. These values, which with the exception of stomach contents display very small scattering within the

[1] See also Chapters 8 and 9.

same group, are generally lower for the germ-free animals. The germ-free small intestine and caecum are isotonic or slightly hypotonic to blood plasma, while in conventional controls, both these values are hypertonic.

TABLE IV

Osmolarity[a] of intestinal contents[b] of germ-free and conventional rats[c]
(m Osm/kg H_2O)
Preliminary results

	Stomach		Small intestine		Caecum		Large intestine	
	M	S.D.	M	S.D.	M	S.D.	M	S.D.
Germ-free	554	250	310	10	283	6	279	6
Conventional	888	173	369	35	392	36	304	27

[a] Determined with Osmometer (freezing point depression), Advanced Instruments, Newton Highlands, Massachusetts, U.S.A.

[b] Samples were weighed, diluted and mixed (distilled water 1:1 ratio), centrifuged 5 min (clinical centrifuge), supernatant pipetted off and refrigerated. From sacrifice to refrigeration of sample 25–30 min elapsed.

[c] Fisher Rats (Charles River), all males, 8–9 weeks old, 180–240 g, corrected for caecal contents, fed on sterilized Purina 5010 C; anaesthesia: urethane 1·5 g/kg body weight.

Average osmolarity of blood plasma: 315.

Number of animals: 6 germ-free, 9 conventional rats.

2. PROTEINS AND RELATED SUBSTANCES

Germ-free animals excrete more faecal nitrogen than conventional controls (Levenson and Tennant, 1963; Luckey, 1963). Combe et al. (1965) and Combe and Sacquet (1966) have found 3–4-fold increase of nitrogen in caecal contents of germ-free rats (when expressed per animal). In the same material the amount of free amino acids was 50–100 times the conventional values and mucoproteins and urea were substantially increased. Lindstedt et al. (1965) demonstrated that germ-free rats excreted more hexosamines than conventional rats, while no difference in nitrogen excretion was found. In germ-free caecal contents, the hexosamines were 30–40 times the conventional values. Combe and Pion (1966) found that among the amino acids of germ-free caecal contents, tyrosine, threonine and methionine, i.e. some of those that are considered characteristic of mucoproteins, were more abundant than in conventional controls. Beaver and Wostmann (1962) found considerably reduced levels of histamine in caecal contents of germ-free rats. In our laboratory, we have recently isolated a pigment fraction in caecal contents of germ-free rats and mice (Gordon and Kokas, 1967) which displays ferritin-like characteristics (in light absorption spectrum, response to anti-ferritin serum, depressant effect on vascular and intestinal smooth muscle and stimulatory effect on spontaneous intestinal villus motility).

3. BILE PIGMENTS

Gustafsson and Swenander-Lanke (1960) found considerably elevated levels of bilirubin in the faeces of germ-free rats and no urobilin in their urine. The amount of faecal bilirubin in these animals was almost the same as the sum of bilirubin and urobilins that are excreted by conventional animals on a daily basis.

4. LIPIDS AND STEROLS

Evrard *et al.* (1964, 1965) found that faecal fatty acids excreted by germ-free rats were largely the unsaturated type and their output was slightly reduced. Cyclical and branched chain fatty acids (which are commonly attributed to bacterial synthesis) were missing in germ-free animals. Gustafsson *et al.* (1957) found that the half life of orally administered labelled taurocholic acid was 11·4 days in germ-free rats compared with 2 days in conventional controls. Germ-free rats maintained on a steroid-free diet excreted in their faeces considerable amounts of neutral sterols, primarily cholesterol. The greater quantities of cholesterol and cholesterol precursors in the germ-free gut might explain the elevated levels of serum cholesterol found in these animals (Danielsson and Gustafsson, 1959). In observations made by Wostmann and Wiech (1961) the levels of serum cholesterol in germ-free rats were not significantly raised; however, liver cholesterol was consistently elevated. Wostmann *et al.* (1966) indicated the slowing effect of germ-free conditions on systemic cholesterol metabolism by showing reduced release of $^{14}CO_2$ after administration of ^{14}C-labelled cholesterol.

5. INTESTINAL ENZYMES

In 1959 Borgström *et al.* reported that in faeces of germ-free rats considerable amounts of trypsin (1–6 mg/day) and invertase (12–25 units/day) were found while in faeces of conventional controls such activities were not demonstrable. The amylase content in the faeces of germ-free rats was the same or slightly higher than in the conventional counterparts. Lepkovsky *et al.* (1964), working on intestinal contents of germ-free chickens, found occasionally an elevation of proteases (trypsin fraction) and of amylases in the contents of the caeca, colon and cloaca. The reverse appeared to be the case with lipase activity. Dahlquist *et al.* (1965), working with homogenates prepared from the combination of caecal wall and contents of germ-free rats, found ten times more maltase and 6-bromo-2-naphthyl-alpha-glucosidase than in conventional animals. Other disaccharidases, glucosidases and galactosidases showed little change from conventional values. Reddy and Wostmann (1966), working with homogenates of the wall of the small intestine of adult germ-free rats, found distinctly higher disaccharidase levels (maltase, invertase, lactase, trehalose and cellobiase) than in conventional controls. Recently we have found considerably elevated levels of a substance in the caecum of germ-free rats and mice that affects smooth muscle *in vitro* and cardiovascular function *in vivo* (Gordon, 1965b). On further scrutiny, it was found (Gordon, 1967) that this substance displays the properties of faecal kallikrein and is capable of releasing a kinin from a suitable precursor. The released kinin appeared similar to kallidin or to bradykinin.

6. TOXICITY OF GERM-FREE CAECAL CONTENTS

We have observed (Gordon, 1965b) that caecal contents of germ-free rats and mice are toxic on intraperitoneal injection. Comparable doses of caecal contents from conventional controls are less toxic. The germ-free preparations (caecal supernatant), when lethal, killed the recipient mice (germ-free or conventional) in 30 min to 6 hr. It was calculated that a germ-free mouse contains in its caecum 5–8 times its own LD_{100}. Dr. William Antopol, from the Beth Israel Hospital in New York, who has kindly examined some of the mice injected with germ-free caecal contents, found haemorrhages in the anterior abdominal wall, intensive vasodilatation of the mesenteric blood vessels and noticable pleural effusion.

III. Some Systemic Characteristics

Desplaces *et al.* (1963) have shown that metabolic rate and fixation of radioactive iodine are significantly lower in germ-free rats in comparison to conventional controls. Matsuzawa and Wilson (1965), working with the effects of bacterial endotoxin, reported slightly reduced oxygen partial pressures in some tissues (subcutaneous, liver) of germ-free mice. The preliminary findings of J. Keene in our laboratory (unpublished) suggest that in an atmosphere of reduced oxygen partial pressure, the survival of germ-free mice is perceptibly smaller than in conventional controls. Her data are illustrated in Fig. 4. Germ-free animals were found to show a mild degree of haemoconcentration and reduction of total blood volume (Gordon *et al.*, 1963; Laroche *et al.*, 1964), although we have not consistently found the difference significant in successive animal groups. Cardiac output was reduced in germ-free rats by over 25% in comparison to conventional controls (Gordon *et al.*, 1963). Regional blood flow determinations carried out in collaboration with Bruckner-Kardoss and

Wostmann indicated greatly reduced values for germ-free rats, mainly in organs that are normally in close contact with the microbial flora (the skin, lungs, digestive tract and the liver); the heart also appeared to be involved. More remote organs, e.g. the kidneys, spleen and thymus, showed only smaller differences in blood flow between germ-free and conventional animals. An illustration of these results is given in Table V. Recently, in collaboration with B. Kelentey (unpublished) the effect of

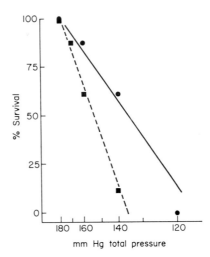

FIG. 4 Survival of mice in decompression experiment. Mice were the same kind as given in Table II (in this instance 50% males and 50% females were used). Total number of animals: 32 germ-free, 32 isolator-conventional mice (8 animals per group). Desiccator-type, sterilized decompression chambers were used within isolators maintaining germ-free or conventional status and satisfactory air flow (20% O_2 concentration, periodically tested with Beckman Oxygen Analyser). Decompression levels were reached uniformly in 1 hr; animals remained at given levels for another hr. All deaths occurred within the second 1-hr period. All observations were made on the same day, in the same room. Degree of decompression was measured with a mercury manometer. Germ-free - - - -; conventional ——.

various vaso- and musculoactive drugs has been tested on blood pressure, respiration and intestinal motility and tone in the intact germ-free and conventional rat. The series of drugs intravenously injected included adrenergic stimulants and blockers (adrenaline, noradrenaline, iso-prenaline, phentolamine and propranolol), cholinergic stimulants and blockers (acetylcholine, atropine), amines and blockers (histamine and

5-HT, tripelennamine and lysergic acid diethylamide) and bradykinin. Working with minimum doses that can elicit responses, we found no essential differences between germ-free animals and their conventional controls, with the exception of histamine and bradykinin where the germ-free rats seemed more sensitive than their conventional counterparts. In earlier work, Gustafsson *et al.* (1957), investigating histamine in the whole body and in the urine of rats, found some differences between germ-free and conventional groups of animals. Distribution and total content appeared to be the same in both groups. In male rats all, or almost

TABLE V

Regional blood flow[a] in germ-free and conventional rats[b]
(whole organ) (ml/min/kg body weight)

Organ	Germ-free		Conventional	
	M	S.D.	M	S.D.
Skin	13·20	2·46	25·94	1·45
Heart (coronary)	3·66	0·63	4·89	0·67
Lung (arterial)	3·17	0·58	5·27	0·91
Small intestine	11·77	2·58	23·57	4·75
Caecum	2·63	0·40	3·35	1·06
Large intestine	2·26	0·60	3·97	0·85
Liver (arterial)	7·44	1·47	14·51	0·80
Liver (portal)[c]	17·68	2·93	32·58	5·13
Thymus	0·28	0·07	0·34	0·08
Spleen	1·02	0·26	1·69	0·18
Kidneys	21·01	2·02	27·74	3·31

[a] Method: Sapirstein (1958).

[b] Wistar strain, 3–3½ months old, 290–340 g males (body weight corrected for caecal contents) fed sterilized L-462 diet; germ-free: 10 rats, conventional: 9 rats. All animals were starved overnight with free access to water. Pentobarbital sodium anaesthesia (35–45 mg/kg body weight, intraperitoneally).

[c] Calculated by adding values of small intestine, caecum, large intestine and spleen.

all, histamine was found in a conjugated form both in germ-free and conventional animals. The germ-free female rats excreted free and conjugated histamine, while the conventional females excreted histamine almost exclusively in the free form.

Recently, it was found that germ-free rats, caecectomized at weaning, failed to develop the haemoconcentration that appears in unoperated germ-free adult animals (Gordon *et al.*, 1966). Repeating our observations on cardiac output in such caecectomized animals, we found that the germ-free values approached (or at times even reached) those of unoperated conventional controls. These observations have focussed our attention on the role which the distended caecum and the large pool of caecal contents may play in the economy of the germ-free host.

IV. Changes with Progress of Age

The best approach for elucidating the effects of the absence or the presence of the microbial flora on the animal host, with special reference to our present inquiry, is a comparative study of germ-free and conventional animals as they age (Gordon, 1966b). In young and young adult animals (where hitherto virtually all observations have been made), some departures from normality may appear slight and their significance on the host's physiological economy remains unclear or altogether escapes detection. As the animals get older, the outlines of the developing pattern become generally more clearly defined. At the same time, a prolonged study of germ-free animals is technically more difficult and, with the increased chances of accidental contamination, more frustrating. To date we have only conducted a systematic study in this area, where natural death was considered as an endpoint (Gordon et al., 1966a). This study, which was carried out in approximately 300 germ-free and the same number of conventional mice, suggested the following trends. (a) Germ-free animals, as a group, showed a significantly longer life span than conventional controls that are fed on the same sterilized diet. Conventional controls maintained on a non-sterile diet lived longer than the sterile-diet conventionals but not so long as the germ-free group. (b) The average survival time in germ-free males and females was the same or even to the advantage of the males. In conventional controls, as in natural populations, the females showed a longer survival time. (c) The caecal anomalies of germ-free animals gradually worsened with progress of age. The loss of muscle tone spread over more extended areas of the gastrointestinal tract (to the small and large intestine and sometimes to the gall bladder). In this context, it was found that the progress of caecal enlargement was accompanied by a parallel increment in the concentration of caecal kallikrein (Gordon et al., 1966a). (d) The natural death of germ-free animals appeared predominantly associated with their peculiar anomalies of reduced muscle tone of the intestine and the excessive filling of the lower bowel. In conventional controls, inflammatory processes were the main cause of death. These results indicated very clearly the fundamentally different types of noxae to which germ-free and conventional animals were exposed. Between the anomalous bowel function of the germ-free and the flora-induced lesions of the conventional animals, the former appeared to be the lesser evil.

V. Microbial Contamination in Redressing Germ-free Anomalies

All or almost all who have engaged in germ-free experimentation with rodents have considered at some time microbial contamination of the animals as a means of reducing the size of the enlarged caecum. This, if successful, might lead to the discovery of an element or elements of the flora which in normal life assist in maintaining normal bowel function, or which might prove indispensable for sustaining physiological normality of the animal host. This is a challenge hard to resist.

It is known that in rats and mice oral contamination with an inoculum of intestinal contents from conventional animals will result in the prompt establishment of a normal microbial population in the gastrointestinal tract and cause the disappearance of the germ-free anomalies within 2–3 weeks, without seriously endangering the life of the animals (Gordon and Wostmann, 1958). Indeed, it is routine procedure nowadays to transfer germ-free rats and mice from their isolators to the open colony rooms and use these animals and their offspring as genetically closely linked conventional controls.

In the quest for identification of elements of the flora which might remedy the anomalies of germ-free life two approaches may be used. One consists of growing various microbial species from conventional intestinal contents in pure culture and then, after oral inoculation, observing their effects as single contaminants or in various known combinations in germ-free animals. This may be termed the direct approach. Another, indirect, method uses treatment of conventional animals with antibiotics as its first step. Under these circumstances, some antibiotics cause considerable reduction of the intestinal flora population and in succession, germ-free-like characteristics may develop in these treated animals. In this approach, it is assumed that there is a causal relationship between these two events and that the microbial species which is found reduced or missing by the antibiotic treatment will identify the associates which in normal life are responsible for elimination of the "germ-free anomalies".

Without claiming completeness of the list, microbial associates tested with the direct approach and found more or less effective in their ability to reduce various germ-free anomalies were as follows: (i) Reduction of the caecum: in the mouse, *Clostridium difficile*, 2 strains of bacteroides (Skelly *et al.*, 1962); combination of lactobacilli, anaerobic streptococci and organisms of the bacteroides group (Schaedler *et al.*, 1965); in the rat, streptococcus sp. (Hudson and Luckey, 1964); *Lactobacillus casei* (Gordon *et al.*, 1966b). Transient caecal reduction in the rat following monocontamination with *Salmonella typhimurium* was first observed by B. S. Wostmann (personal communication) and later elaborated by Wiseman and Gordon (1965). (ii) Reduction of bilirubin to urobilins in the intestine of germ-free rats: *Clostridium welchii* type A and an intensified effect after superinfection with *E. coli* G 14 (Gustafsson and Swenander-Lanke, 1960). It must be emphasized that the mono- and polycontaminants used altered the characteristics of germ-free animals but there is no evidence that these gnotobiotic animals fully resembled conventional rats and mice.

The indirect approach has been tried using procaine penicillin administered orally to chickens (Lev *et al.*, 1957; Gordon *et al.*, 1957; Wagner and Wostmann, 1959) and rats (Gordon *et al.*, 1966b). Following this treatment, the reduction of *Clostridium welchii* was observed in the gut. Yet, in germ-free animals, this bacterium as monocontaminant had virtually no effect in the elimination of germ-free anomalies (although it did stimulate cellular and humoral defensive elements). Coates *et al.* (1963) showed that penicillin did not increase the growth of germ-free chickens. Heggeness (1959) reported that rats orally treated with chloramphenicol or neomycin had greatly distended caeca, predominantly due to increased water content. Meynell (1963) observed in caecal contents of conventional mice treated orally with streptomycin, increased moisture content, more positive oxidation-reduction potential and elevated pH levels in an experiment of a few days' duration. In these circumstances, large numbers of fusiforms, few lactobacilli, coliform organisms, Gram-positive cocci, bacilli (including spore-formers) that are present in caecal contents of untreated animals had disappeared and only scanty Gram-postive bacilli (presumably arising from spores) remained.

VI. Comments and Conclusions

Summing up, it appears that the germ-free animal is unlike its conventional counterpart (apart from lack of development of its defence mechanisms) primarily on two issues. First, a reduction or depression in a number of physiological standards: lower metabolic rate, cardiac output and regional blood flow which may lead to hypoxia in some tissues; reduced muscle tone and water uptake in the lower bowel. Second, accumulation in the intestinal contents of certain metabolites which apparently result from reduced degradation processes under germ-free conditions: excess of various protein components, digestive and other enzymes, bile pigments and sterols. Some of these agents appear to be capable of exerting pronounced biological effects. External causes other than those resulting directly or indirectly from the missing microbial flora do not appear to enter into this picture.

Virtually nothing is known about the mechanism of the decreased oxygen transport in germ-free life. This cannot result from a change in the lungs. As judged by histological criteria, the integrity of the respiratory surface in germ-free animals is much better preserved than in conventional controls (Gordon, 1965a). It has been suggested that the constant mild inflammation which characterizes the microbially exposed organs of conventional animals and which is practically absent in germ-free conditions will also decrease the oxygen demand in these animals (Gordon et al., 1963). On learning recently that the reduced cardiac output, and perhaps also other anomalies of oxygen transport in germ-free animals, may be eliminated by caecectomy, it becomes clear that the presence of the distended caecum and the large pool of caecal contents with its excess of biologically active substances per se may also be implicated in these phenomena.

The reduced muscle tone of the lower bowel in germ-free animals was commonly thought to be the simple result of the absence of flora-stimulation or, more precisely, of missing pressor substances in the gut that are normally supplied by the flora. The findings of profound changes in the autonomic nervous system in this area (Dupont et al., 1965; Jervis and Biggers, 1964) suggested, for the first time, a more complex mechanism. In terms of responsiveness, the smooth muscle of the caecum appeared to be unimpaired as indicated by the prompt contraction to pilocarpine observed by M. Wagner (personal communication). The possibility that bacterial products may play a role in the maintenance of normal caecal tone was shown by Th. E. Staley (personal communication) who found that bacterium-free filtrates prepared from conventional rat caecal contents will cause a clear-cut contraction of germ-free caecal muscle strips in vitro, whereas similar strips from conventional control rats displayed little or insignificant response. The reduction of the size of the caecum and the actual shrinking (reduction in weight) of the caecal sac which were observed in germ-free animals in vivo after contamination with a normal flora inoculum (Gordon and Wostmann, 1958) seemed to be based only in part on contraction or increased tone of the smooth muscle cells. Histological preparations made from this tissue suggested a reduction of the cytoplasmic component of muscle cells, concomitantly with destruction of old and regeneration of new muscle elements. The process is similar to that in a uterus undergoing post-partum involution.

The inhibition of water transport in the germ-free caecum and its possible mechanism are discussed in Chapter 7. The indication that water passes from the blood compartment into the caecal lumen is suggested by the following observations. (a) The haemoconcentration found in germ-free rats decreases after caecectomy. (b) The intestinal contents suddenly increase in liquidity in their passage from the lower ileum into the caecum. (c) The caecal contents of germ-free rats are isotonic with the blood plasma (in conventional controls the caecal contents are slightly hyptertonic). (d) The kallikrein found in the germ-free caecum is similar to plasma-kallikrein and not to pancreatic kallikrein (Gordon, 1967) (the former is blocked by soya bean trypsin inhibitor, the latter is not; Webster and Pierce, 1963). In conventional controls caecal kallikrein more closely resembled the pancreatic enzyme. In general, the greater thirst and the constant mild diarrhoea observed in germ-free animals also suggest impaired intestinal water transport. It appears that the caecum in rodents has a unique part to play in maintaining water balance and in this respect is distinct from other parts of the gastrointestinal tract. This is suggested also by the post-caecectomy state of germ-free animals, where the anomalies presently discussed are, to a large extent, permanently eliminated.

Reports on reduced catabolic activity and the excess of various metabolites in the germ-free gut appeared initially of little consequence, or even advantageous to the host. The recent detection of considerable accumulation of more potent metabolites has changed our view. The source of the kallikrein present in the germ-free gut, as previously suggested, is endogenous. This enzyme, together with kallidin or bradykinin released by it, is known to cause considerable depression of the musculature in the microcirculatory bed and consequently, an increase in the lymph filtration from the blood capillaries into the tissue (Werle, 1934; Fox et al., 1961; Schachter, 1963). Recently, it has been reported that bradykinin has a pronounced depressor effect on the smooth muscle of the lower bowel (Fishlock, 1966). Conditions of hypoxia are known to cause an increased release of kallikrein (demonstrated for salivary kallikrein by Sallay, 1956). The other active substance that has been isolated from germ-free intestinal contents and that may be implicated in the anomalies of these animals is the agent displaying ferritin-like characteristics. Such substances in gut contents may originate from the intestinal epithelium which contains apoferritin, the carrier-protein for the mucosal transport of iron. It has been shown that by desquamation of epithelial cells considerable quantities of ferritin-apoferritin reach the intestinal lumen (Crosby, 1963; Conrad et al., 1964). The reduced form of ferritin (or of apoferritin) is known to be a potent depressant of smooth muscle (Mazur and Shorr, 1948). This form of ferritin is known also to have antidiuretic action which may be eliminated by transection of the hypophyseal stalk (Shorr, 1955). Bauer et al. (1966) could not demonstrate circulating antibodies against ferritin in germ-free rats. Reddy et al. (1965) showed a considerable increment of ferritin iron in the liver of these animals. Thus, it appears that in addition to missing microbial pressor effects, we must consider an excess of depressor substances and their effects in the etiology of reduced smooth muscle tone and possibly in other anomalies of germ-free life. The agents involved may act directly or indirectly or both by way of nervous, humoral or immunological mechanisms. The role of the microbial flora in the inactivation of these agents is clearly implicated.

The worsening of the germ-free anomalies with progress of age is of theoretical and practical interest. It is possible that some forms of impaired bowel function known from clinical work (loss of tone, diarrhoea) may develop on the basis of shifts of the intestinal flora (aided also by antibiotic treatment) and via the mechanisms suggested in this article. Actually, microorganisms whose preponderance in the intestine has been implicated in bowel dysfunction, e.g. *E. coli* in diarrhoeas (Hardy, 1956), putrefactive clostridia in old humans (Orla-Jensen *et al.*, 1949), appear to belong to groups whose representatives as monocontaminants were ineffective in the elimination of caecal anomalies in germ-free animals. Another point of our research in aging germ-free animals has supported the concept that the greater survival of females in conventional populations may be associated with their greater resistance to infections (Wheater and Hurst, 1962). In germ-free conditions this advantage becomes ineffective, hence the disappearance of the female's traditional prerogative to a longer life span under conventional circumstances. The indication that germ-free females may *de facto* have a shorter life span than germ-free males (mean age for females $22 \cdot 7$, for males $24 \cdot 1$ months; statistically significant at $p = 0 \cdot 04$) could be explained by the larger caeca and by the more severe forms of bowel dysfunction which have been seen at natural death in germ-free females (Gordon *et al.*, 1966a).

The knowledge of the mechanisms underlying the anomalies, and their control and elimination while maintaining the germ-free status of the animal, are of prime importance to the student of host-contaminant relationships. To date few methods of control have been tried, with very little success. Among these are feeding of dead bacteria, supplementation of the diet with inhibitors of musculo-depressant and other substances (soya bean trypsin inhibitors to counteract the excess of kallikrein; the binding resin Cholestyramine[1] for cholic acids). An attempt was also made to remedy hypoxia by increasing the oxygen concentration of the inhaled air within the isolators (40% O_2 60% N_2 up to 14 days). None of these methods had an appreciable effect in reducing the size of the enlarged caecum of germ-free animals when used individually. Combinations of such agents will have to be tried. Most successful to date appears to be

[1] Merck, Sharp and Dohme Research Lab., Rahway, N.J.

caecectomy performed in the post-weaning period which, as mentioned, permanently eliminates the large pool of caecal contents.

The search for a viable flora element, which in the form of a known contaminant is capable of eliminating the anomalies of germ-free animals, may lead to the discovery of our hypothetical synergistic microbial commensal which is indispensable for normal life. This, when achieved, will represent the first step toward rational flora control, which aims at the preservation of the synergists and the elimination of other, antagonistic, elements of the flora. This, however, will not solve the problem of the anomalies of germ-free animals which, in order to fulfil their purpose, must remain germ-free.

Returning to the original question of this article, it is clear that the germ-free animal is not normal, neither when compared with many "normal standards" of conventional controls, nor when considering "anatomical and physiological excellence" in various detail, as it is known in conventional life. This statement needs restriction in the sense that it refers primarily to areas or organs of the host which in normal life are in intimate association with the microbial flora. Other, less closely associated areas may not be so much involved; however, *a priori*, one should not take this for granted.

Although germ-free rodents with their monstrous caeca, dirty tails and great complexity of questions have meant disappointment to many an impatient prospective user of the "clean test tube animals", they have shown us, for the first time, what happens when, free of commensals, they are left to their own resources. This is a very important point of departure.

REFERENCES

Abrams, G. D., Bauer, H. and Sprinz, H. (1963). *Lab. Invest.* **12**, 355–364.

Bauer, H., Paronetto, F., Burns, W. A. and Einheber, A. (1966). *J. exp. Med.* **123**, 1013–1024.

Beaver, M. H. and Wostmann, B. S. (1962). *Br. J. Pharmac. Chemother.* **19**, 385–393.

Borgström, B., Dahlquist, A., Gustafsson, B. E., Lundh, G. and Malmquist, J. (1959). *Proc. Soc. exp. Biol. Med.* **102**, 154–155.

Chalkley, H. W. (1943–1944). *J. natn. Cancer Inst.* **4**, 47–53.

Coates, M. E., Fuller, R., Harrison, G. F., Lev, M. and Suffolk, S. F. (1963). *Br. J. Nutr.* **17**, 141–150.

Combe, E., Penot, E., Charlier, H. and Sacquet, E. (1965). *Ann. Biol. anim. Bioch. Biophys.* **5**, 189–206.

Combe, E. and Pion, R. (1966). *Ann. Biol. anim. Bioch. Biophys.* **6**, 255–259.

Combe, E. and Sacquet, E. (1966). *C.r. hebd. Séanc. Acad. Sci., Paris* Sect. D **262**, 685–688.

Conrad, M. E., Weintraub, L. R. and Crosby, W. H. (1964). *J. clin. Invest.* **43**, 963–974.

Crosby, W. H. (1963). *Blood.* **22**, 441–449.

Dahlquist, A., Bull, B. and Gustafsson, B. E. (1965). *Archs Biochem. Biophys.* **109**, 150–158.

Danielsson, H. and Gustafsson, B. E. (1959). *Archs Biochem. Biophys.* **83**, 482–485.

Desplaces, A., Zagury, D. and Sacquet, E. (1963). *C.r. hebd. Séanc. Acad. Sci., Paris* **257**, 756–758.

Dupont, J. R., Jervis, H. R. and Sprinz, H. (1965). *J. comp. Neurol.* **125**, 11–18.

Evrard, E., Hoet, P. P., Eyssen, H., Charlier, H. and Sacquet, E. (1964). *Br. J. exp. Pathol.* **45**, 409–414.

Evrard, E., Sacquet, E., Raibaud, P., Charlier, H., Dickinson, A., Eyssen, H. and Hoet, P. P. (1965). *Ernährungsforschung* **10**, 257–263.

Fishlock, D. J. (1966). *Nature, Lond.* **212**, 1533–1535.

Fox, R. H., Goldsmith, R., Kidd, D. J. and Lewis, G. P. (1961). *J. Physiol.* **157**, 589–602.

Gordon, H. A. (1965a). *Triangle.* **7**, 108–121.

Gordon, H. A. (1965b). *Nature, Lond.* **205**, 571–572.

Gordon, H. A. (1966a). *Abstr. Int. Congr. Geront., VII* Vienna, pp. 50–51.

Gordon, H. A. (1966b). *In* "Perspectives in Experimental Gerontology" (N. Shock, ed.), pp. 295–307. Charles C. Thomas, Springfield, Ill.

Gordon, H. A. (1967). *Ann. N.Y. Acad. Sci.* **147**, 83–106.

Gordon, H. A. and Bruckner-Kardoss, E. (1961a). *Acta anat.* **44**, 210–225.

Gordon, H. A. and Bruckner-Kardoss, E. (1961b). *Am. J. Physiol.* **201**, 175–178.

Gordon, H. A., Bruckner-Kardoss, E. and Wostmann, B. S. (1966a). *J. Gerontol.* **21**, 380–387.

Gordon, H. A., Bruckner-Kardoss, E., Staley, Th.E., Wagner, M. and Wostmann, B. S. (1966b). *Acta. anat.* **64**, 367–389.

Gordon, H. A. and Kokas, E. (1967). *Fedn Proc. Fedn Am. Socs exp. Biol.* **26**, 383.

Gordon, H. A., Wagner, M. and Wostmann, B. S. (1958). *Antibiotics A.,* **1957–1958**, 248–255.

Gordon, H. A. and Wostmann, B. S. (1958). *Int. Congr. Microbiol. VII,* Stockholm, pp. 336–339.

Gordon, H. A., Wostmann, B. S. and Bruckner-Kardoss, E. (1963). *Proc. Soc. exp. Biol. Med.* **114**, 301–304.

Gustafsson, B. E., Bergström, S., Lindstedt, S. and Norman, A. (1957). *Proc. Soc. exp. Biol. Med.* **94**, 467–471.

Gustafsson, B. E., Kahlson, G. and Rosengren, E. (1957). *Acta physiol. scand.* **41**, 217–228.

Gustafsson, B. E. and Swenander-Lanke, L. (1960). *J. exp. Med.* **112**, 975–981.

Hardy, A. V. (1956). *Ann. N.Y. Acad. Sci.* **66**, 5–13.

Heggeness, F. W. (1959). *J. Nutr.* **68**, 573–582.

Hudson, J. A. and Luckey, T. D. (1964). *Proc. Soc. exp. Biol. Med.* **116**, 628–631.

Jervis, H. R. and Biggers, D. C. (1964). *Anat. Rec.* **148**, 591–595.

Laroche, M. J., Cottart, A., Sacquet, E. and Charlier, H. (1964). *Annls pharm. fr.* **22**, 333–338.

Lepkovsky, S., Wagner, M., Furuta, F., Ozone, K. and Koike, T. (1964). *Poultr. Sci.* **43**, 722–726.

Lev, M., Briggs, C. A. E. and Coates, M. E. (1957). *Br. J. Nutr.* **11**, 364–372.

Levenson, S. M. and Tennant, B. (1963). *Fedn Proc. Fedn Am. Socs exp. Biol.* **22**, 109–119.

Lindstedt, G., Lindstedt, S. and Gustafsson, B. E. (1965). *J. exp. Med.* **121**, 201–213.

Luckey, T. D. (1963). *In* "Germfree Life and Gnotobiology", p. 252, Academic Press, New York and London.

Matsuzawa, T. and Wilson, R. (1965). *Tohoku J. exp. Med.* **85**, 361–364.

Mazur, A. and Shorr, E. (1948). *J. biol. Chem.* **176**, 771–787.

Meynell, G. G. (1963). *Br. J. exp. Pathol.* **44**, 209–219.

Nuttal, G. H. F. and Thierfelder, H. (1896). *Hoppe-Seyler's Z. physiol. Chem.* **22**, 62–73.

Orla-Jensen, S., Olsen, E. and Geill, T. (1949). *J. Gerontol.* **4**, 5–15.

Pasteur, L. (1885). *C.r. hebd. Séanc. Acad. Sci., Paris* **100**, 68.

Phillips, A. W., Newcomb, H. R., Smith, J. E. and La Chapelle, R. (1961). *Nature, Lond.* **192**, 380.

Phillips, B. P. and Wolfe, P. A. (1959). *Ann. N.Y. Acad. Sci.* **78**, 308–314.

Pleasants, J. R. (1959). *Ann. N.Y. Acad. Sci.* **78**, 116–126.

Pleasants, J. R. and Wostmann, B. S. (1962). *Proc. Indiana Acad. Sci.* **72**, 87–95.

Reddy, B. S. and Wostmann, B. S. (1966). *Archs Biochem. Biophys.* **113**, 609–616.

Reddy, B. S., Wostmann, B. S. and Pleasants, J. R. (1965). *J. Nutr.* **86**, 159–168.

Reyniers, J. A., Wagner, M., Luckey, T. D. and Gordon, H. A. (1960). *Lobund Rep.* **3**, 7–171.

Sacquet, E. (1959). *Revue fr. Étud. clin. biol.* **4**, 423.

Sallay, C. (1956). *J. dent. Res.* **35**, 840–845.

Sapirstein, L. A. (1958). *Am. J. Physiol.* **193**, 161–168.

Schachter, M. (1963). *Ann. N.Y. Acad. Sci.* **104**, 108–116.

Schaedler, R. W., Dubos, R. and Costello, R. (1965). *J. exp. Med.* **122**, 77–82.

Schottelius, M. (1902). *Arch. Hyg. Bakt.* **42**, 48–70.

Shorr, E. (1955). *In* "Polypeptides which Stimulate Plain Muscle" (J. H. Gaddum, ed.), pp. 120–129. Livingstone Ltd., Edinburgh and London.

Skelly, B. J., Trexler, P. C. and Tanami, J. (1962). *Proc. Soc. exp. Biol. Med.* **110**, 455–458.

Strandberg, K., Sedvall, G., Midtvedt, T. and Gustafsson, B. E. (1966). *Proc. Soc. exp. Biol. Med.* **121**, 699–702.

Vane, J. R. (1957). *Br. J. Pharmac. Chemother.* **12**, 344–349.

Wagner, M. and Wostmann, B. S. (1958–1959). *Antibiotics A.*, **1958–1959**, 1003–1011.

Webster, M. E. and Pierce, J. V. (1963). *Ann. N.Y. Acad. Sci.* **104**, 91–105.

Werle, E. (1934). *Biochem. Z.* **269**, 415–434.

Wheater, D. W. F. and Hurst, E. W. (1962). *J. Path. Bact.* **82**, 117–130.

Wiseman, R. F. and Gordon, H. A. (1965). *Nature, Lond.* **205**, 572–573.

Wostmann, B. S. and Bruckner-Kardoss, E. (1959). *Am. J. Physiol.* **197**, 1345–1346.

Wostmann, B. S. and Bruckner-Kardoss, E. (1966). *Proc. Soc. exp. Biol. Med.* **121**, 1111–1114.

Wostmann, B. S. and Wiech, N. L. (1961). *Am. J. Physiol.* **201**, 1027–1029.

Wostmann, B. S., Wiech, N. L. and Kung, E. (1966). *J. Lipid Res.* **7**, 77–82.

Chapter 7

Intestinal Water Permeability Regulation Involving the Microbial Flora

T. Z. CSÁKY

Department of Pharmacology,
University of Kentucky College of Medicine,
Lexington, Kentucky, U.S.A.

This chapter describes experimental findings and speculates on the possible role of intestinal flora in the water metabolism of higher animals. More precisely, this discussion is limited to one phase—water uptake—since metabolism comprises both intake and output.

I. Water Transport in the Conventional Animal

Unicellular or paucicellular organisms exchange water throughout their entire surfaces. In animals of higher order a special organ, the intestine, serves as the principal, and in some species the exclusive, organ of water intake. In mammals, the body's exposure to the outside world is about 150–200 times greater through the intestine than through the skin. In man, for instance, the surface of the skin is about 1·5 m², whereas the surface of the gastrointestinal tract is estimated to be 200–300 m².

If we limit our attention to the small intestine, it could be considered as a complex membrane dividing two compartments—the lumen and the extracellular compartment of the body. As a rule, biological membranes are permeable to water which traverses in both directions. It could thus be expected that at any given time in the gut there is a simultaneous water excretion (or exsorption) and absorption (or insorption). If the rate of the two processes is equal, no net movement occurs; if the rate is greater in one direction, a net water movement results. It is generally assumed that the movement of water is passive following the "osmotic drag" produced by the movement of ions (Curran and Solomon, 1957; Diamond and Tormey, 1966). Although the active ion transport occurs most likely as a carrier-mediated translocation through the lipid layer, water permeates across hydrophylic pores. The size of these pores is influenced by many factors such as calcium ions and antidiuretic hormone, to mention only two.

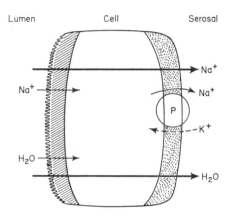

FIG. 1 Principal forces involved in intestinal water transport. Na ions and water enter through the brush-border (lumen-facing membrane). A sodium pump (which may, or may not, also pump potassium simultaneously) is localized in the serosa-facing membrane causing a net transcellular transport of sodium followed by a net water flux.

In the absorption of water from the small intestine, therefore, at least two factors need to be considered: (a) the rate of net ion absorption and (b) the structure of the pores, which in turn governs the passive permeability to water. In the former, sodium ions, which are absorbed by an active transport via a pump localized in the basal (serosal-facing) membrane of the intestinal epithelium, play a primary role. Passive permeability depends on the state of the lumen-facing membrane (brush-border). Factors involved in the intestinal water absorption are schematized in Fig. 1.

It has been repeatedly observed that the direction of net water transport varies in different parts of the intestine. In the duodenum and upper jejunum there is usually a net water excretion; in the middle jejunum and upper ileum there is frequently no net movement, whereas the lower ileum routinely absorbs water. Such typical observation in the chicken is shown in Fig. 2, which illustrates an unpublished experiment by E. Kokas and T. Z. Csáky.

II. Possible Role of the Intestinal Microflora in Water Absorption

A. HYPOTHESIS

It is difficult to find an explanation for variations in water transport solely in the morphological structure or in the distribution of the sodium pump in different parts of the small bowel. The following working hypothesis was therefore advanced. A substance (or substances) is present in the

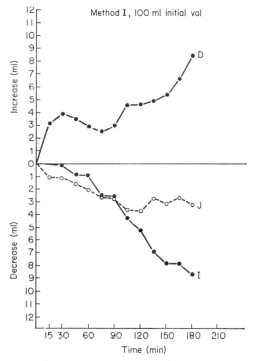

FIG. 2 Water movement in various parts of the intestine of a urethane-anaesthetized chicken. Gut perfused with isosmotic NaCl +phenol red as marker. D =duodenum; J =jejunum; I =ileum. (E. Kokas and T. Z. Csáky, unpublished observations.)

upper part of the small intestine which inhibits water absorption by causing the gut to become less permeable, more "tight" for water. In the lower part of the intestine this substance undergoes destruction, perhaps by intestinal bacteria, and thus enables the mucosa to absorb water.

This hypothesis was tested experimentally in rats and it was found that the duodenum and especially the stomach contain an inhibitor for the absorption of water from the lower ileum. For the sake of brevity, the hypothetical substance(s) is called water-absorption-inhibitor (WAI). It was subsequently found that saliva also inhibits water absorption. The WAI is probably destroyed by intestinal bacteria since the caecal content of germ-free, but not conventional, animals contains it. The ileum of germ-free animals actually absorbs water only after it has been perfused with saline for some time, as if a "wash out" of the inhibitory action is first necessary.

A brief description of the experiments and a short discussion of the possible significance of the present study follow.

B. EXPERIMENTAL PROCEDURE

The experiments were conducted in Sprague-Dawley male rats weighing approximately 200 g, anaesthetized with urethane (1·5 g/kg, subcutaneously). Loops of the lower ileum, 8–10 cm long, were cannulated and perfused by recirculation using either of the two standard procedures of this laboratory. One of the procedures (Method I) requires 25–50 ml fluid for the recirculation; the other (Method II) utilizes smaller volumes, 6–10 ml (Csáky and Ho, 1965). The basic perfusing fluid was an isosmotic NaCl to which human haemoglobin was added as a marker. Small samples were withdrawn from time to time and diluted to 5 ml with distilled water. One drop of conc. HCl was then added, the sample was well mixed and after standing at room temperature for 30 min it was read in the colorimeter at approximately 540 mμ. Assuming that haemoglobin is an ideal marker, viz. it is neither absorbed from the lumen nor adsorbed on to the brush-border, the water absorption could be calculated. The latter is given in the following figures either as percentage of the offered fluid or simply by indicating the increase of the haemoglobin concentration.

In the first set of experiments the duodenum was perfused for 1 hr with 10 ml of saline; this perfusate was then added to a solution perfusing the lower ileum. In the presence of the duodenal washing there was occasionally a somewhat slower rate of water absorption from the ileum.

Next, the content of the stomach was examined as follows. Starved rats were anaesthetized with ether, the pylorus was ligated, the wound was closed and the animals were allowed to wake up. No food or water was offered for approximately 6 to 8 hr, after which time the animal was killed, the stomach removed and emptied and its content neutralized and adjusted to be approximately isosmotic. The adjusted stomach content was

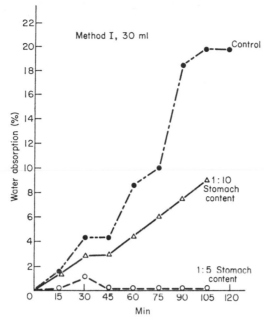

FIG. 3 Effect of rat stomach content on the absorption of water from the ileum of the rat.

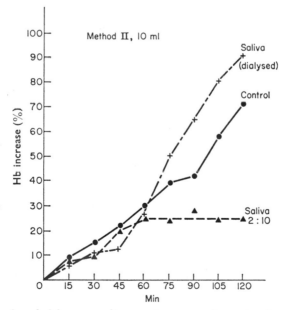

FIG. 4 Effect of pooled human saliva on the water absorption from the ileum of the rat. The inhibitory effect is lost after dialysis.

then added to the saline perfusing the ileum. Figure 3 indicates that the stomach content strongly inhibited the absorption of water.

The stomach content, obtained as previously described, contains a large amount of saliva which the rat swallows, in addition to the secretion of the stomach. In order to distinguish which of the two components contained WAI, a pure neutralized stomach secretion of a dog was first examined; it did not inhibit water absorption. Next, pooled human saliva was added to the perfusate. Figure 4 shows the inhibitory action.

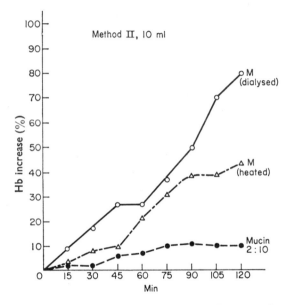

FIG. 5 Effect of the crude "mucin" (M) prepared from human saliva on the water absorption from the ileum of the rat. Heating or dialysis diminishes the inhibitory effect.

The kallikrein content of the saliva is high (Frey *et al.*, 1950) and it can be anticipated that it also contains or liberates kallidin or bradykinin, which in turn may influence water transport. In a set of experiments, synthetic bradykinin was therefore added to the ileum perfusate. Even very large concentrations (up to 2 mg/l) were without effect on the water absorption.

Human saliva was subjected to a crude fractionation. At pH 4 the acid mucoproteins and mucopolysaccharides (mucin) were precipitated. The centrifuged precipitate was redissolved at neutral pH. As shown in Fig. 5, the inhibitory effect is recovered in the crude "mucin" fraction. The salivary WAI is heat labile. It is destroyed if heated at 80° C for 15 min

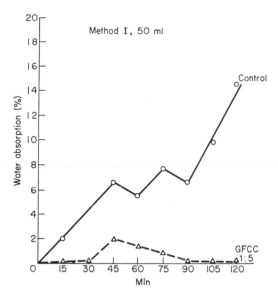

FIG. 6 Effect of caecal content of germ-free mice (GFCC) on the water absorption from the ileum of the rat.

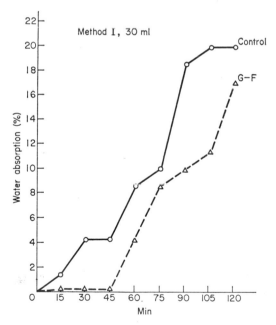

FIG. 7 Absorption of water from the ileum of a conventional rat (Control) and a germ-free rat (GF).

(Fig. 5). Dialysis overnight at 0° C against saline decreases its inhibitory action (Figs 4 and 5).

If human saliva and an appropriate nutrient solution mixture are incubated at 37° C with rat faeces, the WAI action is diminished but not destroyed, even though there is a visible growth of bacteria. Nevertheless, the content of the caecum of germ-free mice (kindly supplied by Dr. H. A. Gordon of this laboratory) was inhibitory to water absorption (Fig. 6). If we assume that the caecum of germ-free animals harbors the WAI, as it is not destroyed in the gut by bacteria, we would anticipate that in the ileum of axenic animals the water absorption is normally inhibited. Figure 7 shows a typical experiment in which the ileum of a germ-free rat (also supplied by Dr. Gordon) was perfused. Curiously, there is no difference in the ultimate rate of water absorption in the normal and germ-free rat gut except that absorption does not start in the latter for 30–60 min—as if something were there which initially inhibited it but which, after sufficient washing, was removed. It is interesting to note that identical results were obtained regardless of whether the intestine of the germ-free rat was perfused under rigidly sterile conditions or not.

C. DISCUSSION

The above outlined experiments show that the presence in the intestinal lumen of contents (most prominently saliva) of the higher alimentary tract inhibits the absorption of water from the lower part of the small intestine. It can be assumed that the presence of the inhibitor is responsible for the fact that no net water absorption is detected in the upper and sometimes the middle part of the intestine. In the normal intestine the inhibitor is gradually destroyed, whereas in the germ-free intestine it is present even in the caecum. Consequently, it can be assumed that the normal intestinal flora is primarily responsible for the elimination of the salivary inhibitory substance. This would represent a hitherto unrecognized function of the intestinal flora in the regulation of the body's water metabolism.

Water balance is the result of a dynamic equilibrium activated by the correlated function of many factors. The elimination of any one of these factors may produce a new, abnormal water balance. As one would anticipate that in the absence of intestinal flora the entire length of the small intestine is exposed to the effect of the water transport inhibitor, it is possible that this continuous inhibitory effect may lead to increase of the water content of faeces, and increase of volume of the intestinal content, causing a dilatation of part of the intestine, particularly if the tone of the muscle is already weakened. *Ceteris paribus*, the decreased intestinal water transport will necessarily lead to a diminished water concentration of the blood and tissues, to a haemoconcentration accompanied by polydypsia. Indeed, a large bowel, hydropenia, haemoconcentration and polydypsia are all found in the germ-free animal (see Chapter 6) and all could be related to one single cause, namely the lack of bacterial destruction of the inhibitor of the intestinal water absorption.

It should be emphasized that the present study was aimed solely at introducing the concept of the possible role of intestinal bacteria in water metabolism. The experimental data seem to justify the above outlined working hypothesis. Considerably more work is needed, however, to clarify the nature of the water absorption inhibitor, the mechanism of its action and the way in which intestinal bacteria are involved in its destruction before a firm theory can be developed.

REFERENCES

Csáky, T. Z. and Ho, P. M. (1965). *Proc. Soc. exp. Biol. Med.* **120**, 403–408.

Curran, P. F. and Solomon, A. K. (1957). *J. gen. Physiol.* **41**, 143.

Diamond, J. M. and Tormey, J. McD. (1966)., *Fedn Proc. Fedn Am. Socs exp. Biol.* **25**, 1458–1463.

Frey, E. K., Kraut, H. and Werle, E. (1950). "Kallikrein (Padutin)." F. Enke, Stuttgart.

Chapter 8

Nutrition and Metabolism

M. E. COATES

National Institute for Research in Dairying, (University of Reading)
Shinfield, Reading, England

The microbial population of the gastrointestinal tract of animals is so numerous and complex that its influence on the nutrition of the host is likely to be considerable. The part played by microorganisms in the accepted nutritional processes has long been a matter of speculation. Gnotobiotes offer a means of studying the course of digestion, absorption and metabolism of nutrients in the absence of microorganisms and the consequent modification of those processes on association with one or more components of the conventional gut flora.

The interpretation of studies of comparative nutrition between germ-free and contaminated animals is not simple. For instance, the absorptive capacity of the intestine depends on villus structure, surface area and integrity of the mucosa. In view of the differences in intestinal morphology between germ-free and conventional animals (cf. Chapters 5 and 6), any observed effect of association with microorganisms may simply be part of the general reaction to bacterial challenge rather than a direct response to a specific organism. The behaviour of any individual component of the conventional microflora is conditioned by the presence of others with which symbiosis or some other relationship has been established and its metabolic and invasive properties may not be the same when it is introduced alone into an otherwise germ-free gut. Even when a particular microbial influence on the end-products of metabolism of a nutrient has been demonstrated, its significance in the overall nutrition of the host has still to be determined.

With these limitations in mind, information on the role of the gut flora in the processes of nutrition is steadily accumulating. So far, the lipids have received most study and are dealt with separately in Chapter 9. Information regarding other dietary components is reviewed below.

I. Major Dietary Components

A. CARBOHYDRATES

Little is known regarding the comparative utilization of carbohydrates by germ-free and conventional animals but some of the enzymes concerned in their digestion have been investigated, mainly with a view to determining their origin. In the rat small intestine, Dahlquist *et al.* (1965) found 6-bromo-2-naphthyl-glucosidase, β-galactosidase and other disaccharidase activities to be much the same in germ-free and conventional animals and concluded therefore that none of the enzymes concerned was of bacterial origin. In the caecum, the much higher concentration of maltase and 6-bromo-2-naphthyl-α-glucosidase in the germ-free rats suggests that these enzymes may be destroyed by bacteria in the conventional gut. Since the pH optima of 6-bromo-2-naphthyl-α-glucosidase were different in the caecum and small intestine of conventional but not germ-free rats, it seems probable that some of the caecal enzyme in conventional rats is a bacterial product.

Reddy and Wostmann (1966) found that the pattern of development of intestinal disaccharidases in young growing rats was the same in germ-free and conventional conditions; at weaning a decrease in lactase and cellobiase activity was associated with increased maltase, invertase and trehalase activities in both groups. In adult rats all disaccharidase activities in the small intestine were lower in conventional animals, a finding that is somewhat at variance with that of the earlier authors but nonetheless confirms their conclusion that the disaccharidases of the rat small intestine are not of microbial origin.

Amylase activities at different sites in the gastrointestinal tract of chicks (Lepkovsky et al., 1964) and rats (Lepkovsky et al., 1966) were similar in germ-free and conventional environments. Amylase was the least stable of the pancreatic enzymes and very little activity survived as far as the caecum.

There is contradictory evidence regarding the regulation of glucose metabolism in germ-free and conventional rats. In glucose tolerance tests, Desplaces et al. (1965) observed no difference in response between germ-free and conventional animals, but Wiech and Hamilton (1966) reported slower removal of glucose from the plasma of germ-free compared with conventional rats and a higher fasting level of plasma glucose in the germ-free animals. The hypoglycaemic effect of some antidiabetic sulphonamides was independent of microbial action (Zagury et al., 1962).

B. PROTEINS

Studies of the proteases in the pancreas and intestinal contents of chicks (Lepkovsky et al., 1964) and rats (Lepkovsky et al., 1966) indicated little difference between germ-free and conventional animals, apart from a tendency to lower values in the germ-free caecal contents of chicks. In the caecal contents of both species the proportion of non-protein nitrogen was much higher in germ-free conditions, although the total nitrogen was similar in both environments.

Other studies with rats (Levenson and Tennant, 1963; Evrard et al., 1964) have shown a greater faecal nitrogen excretion in germ-free compared with conventional animals. Urinary nitrogen output was similar in both environments (Levenson and Tennant, 1963). At least part of the faecal nitrogen from conventional animals is of bacterial origin; the difference between germ-free and conventional animals in their excretion of nitrogen from non-bacterial sources must therefore have been even greater than the difference in their excretion of total nitrogen. In germ-free faeces, Levenson and Tennant (1963) reported a higher proportion of filterable nitrogen which was not precipitable by trichloracetic acid and Evrard et al. (1964) found that urea, which was absent from the conventional samples, accounted for at least 25% of the faecal nitrogen from germ-free rats.

Unless particular care was taken in these experiments to prevent bacterial action in the conventional faeces during the period of collection, aerobic activity after voiding may have altered the faecal composition to some extent. However, it is clear from analyses of caecal contents (Combe et al., 1965) that differences in the nitrogenous constituents of the digesta exist even before excretion (Table I). In germ-free rats the caeca contained quantities of free amino acids many times higher than in conventional rats. There was an appreciable amount of urea but little ammonia; conversely, conventional caecal contents contained no trace of urea but ten times as much ammonia. In later work with rats given high (22%) or low (7%) protein diets (Combe and Sacquet, 1966) the higher nitrogen content in the caeca of germ-free rats was accounted for by an

TABLE I

Analysis of the soluble nitrogen content of the caeca of germ-free and conventional rats (in mg/100 g fresh contents)

Type of rat	No. of rats	Mean weight Carcass	Mean weight Caecum	Total	Nitrogen analysis As NH_3	As urea	As amino acids
Conventional	5	206	2·7	220	22·3	0	2·7
Germ-free	10	184	14·1	417	2.25	19·5	131·7
	6	198	17·4	454	3·82	26·0	244·9

from Combe et al. (1965).

increase in ethanol-soluble compounds (e.g. free amino acids, urea, peptides) and mucoproteins; the amount of urea increased with the level of dietary protein. Lindstedt et al. (1965) also found considerably more nitrogen in the caeca of germ-free rats, although the total faecal nitrogen did not differ appreciably from that of conventional controls. Total hexosamines, which were taken to be an index of mucous material, were much higher in germ-free caecal contents and faeces.

The nitrogenous constituents of gut contents and excreta originate partly from dietary sources and partly from endogenous nitrogen compounds in desquamated epithelial cells, digestive enzymes, mucus and filterable constituents of the plasma, including urea. The presence of urea in caecal contents and faeces of germ-free but not of conventional rats is compatible with the findings of Levenson et al. (1959) who followed the fate of subcutaneously-injected ^{14}C-urea in rats in both environments. The germ-free animals expired [14]CO_2 in amounts equivalent to only 0·02% of the administered ^{14}C-urea, whereas the controls expired nearly one hundred times as much. The authors concluded that bacterial urease is mainly responsible for the hydrolysis of urea in mammals and hence, in the absence of microorganisms, urea is the true end-product of amino acid catabolism. Contamination of germ-free rats with organisms isolated from the digestive tract of conventional rats showed that a strain of *Staphylococcus pyogenes* and of *Actinobacillus* and several strains of lactobacilli were capable of hydrolysing urea in the caecum (Ducluzeau *et al.*, 1966). The results of Combe *et al.* (1965) are in accord with those of Warren and Newton (1959), who reported a four-fold increase in ammonia in the portal blood of conventional compared with germ-free guinea pigs.

It is clear from the foregoing results that, whatever the source of nitrogen, the end products of its metabolism are modified in the presence of a microflora in the alimentary tract. There is, however, little indication as to how far the endogenous nitrogen excretion is affected, if at all, by microorganisms, nor of the importance of the altered nitrogen metabolism to the general nutrition of the host. In the rat, both questions are complicated by the occurrence of an enlarged caecum in the germ-free animal, which makes quantitative differences in excretion products difficult to interpret.

Preliminary studies with chicks (Miller, 1967) showed no difference between germ-free and conventional birds in their efficiency of protein utilization. Nitrogen balance (nitrogen intake minus nitrogen excretion) was determined in chicks given a nitrogen-free diet or one containing a protein supplement of casein and gelatin. There was a greater total nitrogen excretion by the germ-free birds, although the difference was significant only on the nitrogen-free diet. Nevertheless, the net protein utilization of the casein-gelatin diet, calculated as the ratio of the difference in nitrogen balance to the difference in nitrogen intake on the two diets, was virtually the same in both environments. The protein supplement in these tests was provided in ample amount (26%) and a readily digestible form; it remains to be determined whether proteins supplying a less generous quantity of available amino acids would be equally well utilized in the presence as in the absence of a gut microflora.

II. Vitamins

The well-accepted belief that many of the vitamins are synthesized by microbial fermentation in the alimentary tract has hitherto been largely based on indirect evidence. For instance, in coprophagous animals such as the rat it is difficult to induce a deficiency of vitamin K or some of the B complex; chickens maintained on deep litter or free range derive a significant contribution to their vitamin requirement by ingestion of their excreta. Gnotobiotic studies have now clearly established the ability of gut microorganisms to synthesize vitamins and have thrown some light on the extent of this synthesis and its importance to the host animal.

A. VITAMINS OF THE B COMPLEX

There is considerable evidence that the addition of penicillin to diets lacking thiamine ameliorates the signs of its deficiency in rats. Such findings could be explained in terms of either a greater microbial synthesis or increased absorption of the vitamin in the animals given the antibiotic. Experiments with germ-free rats (Wostmann et al., 1958) support the former suggestion. In a comparison of germ-free and conventional rats given a diet low in thiamine supplemented or not with procaine penicillin, typical effects of thiamine deficiency were observed in all the germ-free animals, whether or not they had received the antibiotic. In corresponding animals reared conventionally, those receiving penicillin grew better and had higher reserves of thiamine than their untreated controls, which developed acute signs of thiamine deficiency. From these results there is a strong indication that dietary penicillin increases microbial synthesis of thiamine in the alimentary tract; the synthesized vitamin is available to the host, although it is not clear whether absorption occurs directly or only after recycling by coprophagy.

The microbial synthesis of thiamine and its importance to the host have been investigated in some detail by Wostmann and his colleagues (Wostmann and Knight, 1961; Wostmann et al., 1962, 1963). After administration by stomach tube of $^{35}SO_4''$ to conventional rats, large amounts of ^{35}S-thiamine were recovered from the caecal and colon contents; little activity was detected in the upper parts of the digestive tract. Comparison of the thiamine levels in the gut contents of germ-free and conventional rats also indicated considerable production of thiamine by the caecal flora in conventional animals. However, surprisingly little of the microbially-synthesized vitamin appeared to be utilized by the rat, even though some coprophagy undoubtedly occurred. In the experiments with $^{35}SO_4''$ little or none of the ^{35}S-labelled thiamine appeared in the heart or liver. On diets low in thiamine, growth of conventional rats was marginally better than that of their germ-free controls but the dietary thiamine requirement for optimal growth was the same in both environments.

Chemical and electrophoretic analysis showed that about one half of the caecal thiamine in conventional rats was readily extractable with saline; the greater part of the remainder was firmly cell-bound and probably unavailable for absorption either directly or even on recycling after coprophagy. Most of the bound thiamine was in the form of an active decarboxylase.

The liver content of thiamine was consistently lower in germ-free than in comparable conventional rats. It is unlikely, in view of the contrary evidence quoted, that the higher liver stores in conventional rats resulted from absorption of microbially-produced thiamine, particularly as the heart content was similar in both environments. Instead, the authors suggest that the thiamine level in the liver might be dependent on metabolic activity. In animals on a low carbohydrate diet, with consequently less demand for thiamine in the form of co-carboxylase, liver levels of thiamine were reduced in both germ-free and conventional rats. In the same way, the absence of bacteria and of the need to detoxify their products would decrease the energy requirement of the liver and probably, therefore, its content of thiamine.

In a study with rats given a diet low in folic acid (Daft *et al.*, 1963) germ-free but not conventional animals began to decline in growth after about two months. They also developed the leucopenia, anaemia and urinary excretion of formiminoglutamic acid that are typical of folic acid deficiency. Monoassociation of the germ-free rats with cultures of *Aerobacter* spp., *Alcaligenes* spp., *Escherichia coli* or *Proteus* spp. protected them against the signs of deficiency. In a further experiment two rats that were maintained germ-free died after about nine weeks with the classic signs of a lack of folic acid, whereas six others monoassociated with *E. coli* were protected against the deficiency. Of these, three were fitted with tail cups to prevent coprophagy and although they did not thrive so well as the uncupped rats, their performance was sufficiently superior to that of the germ-free animals to allow the conclusion that they were able to benefit to some extent from the microbially synthesized vitamin. Whether the folic acid was available to them as a result of direct absorption from the lower gut or through synthetic activity of organisms in the oral cavity is not clear. Miller and Luckey (1963) reached similar conclusions in experiments with chicks given diets low in folic acid. Birds monoassociated with a strain of *E. coli* from conventional chickens had higher haemoglobin values and tissue contents of folic acid than their germ-free counterparts, even though coprophagy was prevented and vitamin production by bacteria growing in the drinking water was shown to be negligible.

Conversely, pantothenic acid, although synthesized in the gut of conventional rats and chicks, appears to be unavailable without coprophagy. In conventional rats given a diet low in pantothenic acid (Daft *et al.*, 1963) protection against deficiency of the vitamin was afforded by supplements of antibiotics or ascorbic acid, but in germ-free rats, or conventional rats fitted with tail cups, such supplements did not alter the course of development of the deficiency. The authors interpret the results as indicative of synthesis of pantothenic acid in the gut of animals given antibiotics or ascorbic acid but failure to absorb the vitamin without coprophagy. Coates et al. (1967) detected negligible amounts of pantothenic acid in the caecal contents of germ-free chicks given a diet devoid of the vitamin, but found appreciable quantities in the caecal contents of similar birds maintained conventionally without access to their excreta. However, since the conventional birds did not grow better or have higher reserves of pantothenic acid in the liver, it was concluded that they did not absorb the microbially-synthesized vitamin. Similar results were obtained for riboflavin and vitamin B_6.

In chicks reared on a purified diet with cyanocobalamin as the only source of vitamin B_{12} (Coates et al., 1963b) significant amounts of the vitamin B_{12}-like factors (pseudo-vitamin B_{12}, factors A, B and probably F) were found in the caecal contents and droppings of conventional birds but were absent from samples from germ-free birds. True vitamin B_{12} was also synthesized by microbial action in the gut (Coates et al., 1968) but was not utilized apparently by birds in which coprophagy was prevented. In germ-free rats on a diet low in vitamin B_{12}, Valencia et al. (1965) found relatively small amounts of cobalamins in the faeces compared with the quantities present in faeces from conventional rats. Monoassociation with a strain of *Veillonella* induced a considerable synthesis as judged by the faecal content of cobalamins but did not increase tissue reserves of vitamin B_{12} to any marked extent.

B. ASCORBIC ACID

In contrast to general experience with the vitamins B, the requirement for vitamin C seems to be lower in germ-free animals. In guinea pigs maintained from birth on a scorbutogenic diet, scurvy developed in germ-free and conventional animals at about five weeks of age. However, on administration of vitamin C the germ-free guinea pigs recovered very much more rapidly (Phillips et al., 1959). In a more extensive study (Levenson et al., 1962) germ-free guinea pigs were compared with "conventionalized" animals delivered germ-free but immediately contaminated by caecal contents from conventional guinea pigs. All received a normal diet for one to two months, after which the ascorbic acid was omitted. After twenty-one days' deprivation, gross lesions of scurvy were observed in the conventionalized guinea pigs but only minor ones in their germ-free counterparts; most of the conventionalized guinea pigs had died after thirty days on the scorbutogenic diet but the germ-free animals survived for between thirty-seven and fifty-seven days. The increased requirement for vitamin C by the animals harbouring a flora may indicate utilization of ascorbic acid by microorganisms at the expense of the host. The authors suggest also that it may reflect an increased demand for the vitamin as part of an overall metabolic response to infection; they cite as a possible analogy the increased requirement for ascorbic acid in the general metabolic disturbance that follows trauma.

C. FAT-SOLUBLE VITAMINS

With the notable exception of vitamin K, there are few indications that utilization of the fat-soluble vitamins is greatly influenced by the flora of the alimentary tract. Higher liver reserves of vitamin A have been found in germ-free compared with conventional chicks given diets of natural or purified ingredients (Thompson and Coates, unpublished); this finding could not be accounted for in terms of a more efficient conversion of carotenoids, since the purified diet contained only vitamin A and none of its precursors. Reports that large doses of vitamin A exacerbate vitamin K deficiency in rats have been confirmed (Wostmann and Knight, 1965) but the suggestion that excessive amounts of vitamin A might interfere with microbial synthesis of vitamin K has been disproved, since the phenomenon occurred with equal severity in the germ-free and conventional rats.

The belief that vitamin K is synthesized by intestinal microorganisms in quantities nutritionally adequate for the host has been substantiated by experiments in germ-free rats. Gustafsson (1959) showed unequivocally that germ-free rats maintained on a diet devoid of vitamin K rapidly developed prolonged prothrombin times and haemorrhages characteristic of its deficiency but conventional rats on the same diet remained in good condition with normal prothrombin times. Administration of vitamin K preparations, or removal to a contaminated environment, brought about a dramatic recovery in the germ-free animals. In later work these findings were repeated and extended with two different strains of rat (Gustafsson et al., 1962). Vitamin K_1 at a dose of 25 mg/kg body weight reversed the hypoprothrombinaemia in the germ-free rats; menadiol tetrasodium diphosphate and menadiol sodium sulphate were also effective, although ten and one hundred times as much, respectively, were necessary. Wostmann et al. (1963) found menadione to have only one-tenth the activity of vitamin K_1 for germ-free rats. On association of germ-free rats with organisms isolated from the oral cavity and faeces of conventional rats, only a strain of E. coli and a sarcina-like micrococcus, which was not identified, effectively reversed the signs of vitamin K deficiency (Gustafsson *et al.*, 1962). The failure of earlier workers (Luckey *et al.*, 1955) to demonstrate vitamin K deficiency in germ-free rats was probably due to traces of the vitamin in the major components of the diet, since in the same laboratory Brambel (1960) was able to produce the typical haemorrhagic syndrome in germ-free rats after the main ingredients of the diet had been extensively extracted with solvents to remove traces of vitamin K.

III. Inorganic Elements

A. IRON AND COPPER

The observation of a hypochromic anaemia in germ-free rabbits (Reddy *et al.*, 1965) indicated the possibility that the metabolism of iron or copper might be influenced by the presence of microorganisms of the alimentary tract. On two steam-sterilized diets that had proved adequate in a conventional environment, germ-free rabbits developed low haemoglobin and haematocrit values. Their plasma levels of iron were low but of copper were within the normal range; the total iron-binding capacity of the plasma was high. Removal of some of the animals to conventional quarters brought about a rapid return to normal. In germ-free rabbits given a soya bean meal diet that contained more iron from natural ingredients, no anaemia was observed. Furthermore germ-free rabbits given either of the original diets recovered from the anaemia when transferred to the soya diet. It was concluded that the availability of iron to the rabbit is reduced in the absence of a gut flora but that the soya meal diet provided iron in a form that was not seriously affected by the germ-free state. In contrast, studies with rats gave only slight indications that iron may be less available in the absence of a gut flora (Reddy *et al.*, 1965b). Differences in storage were observed; the spleen and kidney of germ-free rats contained less, but the liver more, than their conventional counterparts. The relative proportions of haemosiderin and ferritin iron were comparable in both groups. Levels of copper followed the same pattern as those of iron.

A further indication of the probable influence of the gut flora on iron metabolism is the observation of dense foci of an iron-containing pigment in the *tunica propria* in conventional but not germ-free guinea pigs (Geever and Levenson, 1964). It has been shown with rats (Wostmann and Bruckner-Kardoss, 1966) and guinea pigs (Phillips *et al.*, 1959) that the oxidation-reduction potential in caecal contents is more positive in germ-free than in conventional animals. If this is true also for intestinal contents, it is likely that availability of certain nutrients will be affected.

A greater proportion of iron, for instance, may be in the more readily available ferrous form.

B. CALCIUM

A disturbed pattern of calcium metabolism in the germ-free rat is indicated by the observation of a high incidence of urinary calculi which were completely absent from conventional rats (Gustafsson and Norman, 1962). Examination of the urinary electrolyte excretion in the germ-free rats showed a high calcium and citrate but low phosphate concentration, with the levels of magnesium, potassium and sodium very little different from the conventional pattern. The stone-forming tendency disappeared and the urinary electrolyte pattern returned to within the normal range when the rats were associated with the intestinal flora of conventional controls. X-ray diffraction analysis revealed that the stones were composed predominantly of calcium citrate hexahydrate with areas of calcium oxalate dihydrate and, occasionally, basic calcium phosphate (Glas and Gustafsson, 1963). As yet there is no evidence as to whether the high urinary excretion is the consequence of an increased absorption of calcium or of some modification of mineral metabolism in the germ-free state. There is, however, evidence from a small experiment with chickens (Edwards and Boyd, 1963) that calcium uptake is greater in germ-free birds.

C. OTHER ELEMENTS

Measurements of tissue levels of manganese in germ-free and conventional rats gave no indication of differences in its metabolism in the two environments (Reddy et al., 1965a).

Combe et al., (1965) determined sodium and potassium in caecal contents of germ-free and conventional rats and, because the germ-free caeca with their contents were five to six times heavier, expressed the results both in absolute amounts and in terms of concentration. The quantity of sodium per caecum was six to ten times greater in the germ-free rats, although its concentration was similar to that in conventional rats. Conversely, the total amounts of potassium in germ-free and conventional caeca were comparable but its concentration was much lower in the germ-free animals.

IV. Relation of the Gut Microflora to Growth

The demonstration that small dietary supplements of antibiotics and other chemotherapeutic agents increased the growth rate of young animals drew attention to the possibility that certain components of the microflora of the alimentary tract might interfere with an animal's ability to develop its full growth potential. If the hypothesis is tenable, it follows that any dietary constituent capable of eliminating such organisms, or discouraging their establishment, might have a beneficial effect on growth; conversely, poorer growth would be likely on diets that favoured development of undesirable elements in the flora. Investigations have been made with gnotobiotic chicks to test the hypothesis and, where appropriate, to characterize the responsible organisms.

A. ANTIBIOTICS AND CHICK GROWTH

There is ample evidence (Table II) that the growth of germ-free chicks is not improved with dietary supplements of antibiotics. In the experiments of Lev and Forbes (1959), introduction of a strain of *Clostridium*

TABLE II

Effect of dietary supplements of antibiotics on chick growth

Source of data	Type of diet	No. of chicks per treatment	Age in weeks	Mean weight of chicks Basal diet	Diet + antibiotic
Forbes and Park (1959)	Casein/starch	22	4	358	350
	Soya/maize	22	4	349	339
Lev and Forbes (1959)	Casein/starch	14	3	257	254
		18–20	3	240	241
Coates *et al.* (1963a)	Practical chick starter	50	4	300	292

welchii with or without other organisms from the gut of conventional chicks, brought about a depression of growth that was counteracted to some extent by supplements of procaine penicillin in the diet. However, since other workers failed to produce a growth depression in germ-free chicks by association with *Cl. welchii* (Eyssen and De Somer, 1965; National Institute for Research in Dairying, 1966), the question of the role of this organism in the growth depression reversible by penicillin remains unresolved.

On purified diets containing a high proportion of sucrose, Eyssen and De Somer (1963) described a "malabsorption syndrome" in chicks, a condition in which a check in growth was accompanied by a thickening of the gut wall and impaired absorption of fats and carbohydrates. The syndrome was absent from germ-free chicks and could be prevented in conventional birds by inclusion of antibiotics in the diet. Association of the germ-free chicks with strains of organisms from conventional birds, either singly or in combination, failed to reproduce the syndrome exactly, although a small but consistent effect was obtained with a strain of *Streptococcus faecalis* (Eyssen and De Somer, 1965). Similar findings were reported by Huhtanen and Pensack (1964). Simultaneous administration to germ-free chicks of enterococci and a bacteria-free extract of faeces from conventional chicks resulted in a condition very similar to the "malabsorption syndrome" seen in conventional birds (Eyssen and De Somer, 1965). The active principle in the faeces filtrate could not be precisely identified but had all the characteristics of a virus (Eyssen and De Somer, 1967). The authors postulate that the primary mucosal lesion is probably caused by a virus in the faeces filtrate and that subsequent bacterial invasion results in full development of the "malabsorption syndrome".

Although it seems clear that the gut microflora is concerned with the growth depression relieved by dietary antibiotics in conventional chicks, these gnotobiotic studies offer no simple explanation of the phenomenon. It appears likely that the responsible organisms may differ with the diet and management of the birds and that the severity of their effect can be modified according to the conditions prevailing in the digestive tract.

B. UNIDENTIFIED GROWTH FACTORS

The existence of growth factors for chicks in substances such as fish solubles, whey and fermentation residues has frequently been claimed. Failure to characterize the active factor(s) has been due partly to inconsistency in the response of chicks used to assay the crude materials and fractions prepared from them. Barnett and Bird (1956) observed that a lack of growth response to fish solubles often coincided with a failure of antibiotics to promote growth and postulated that components of the gut flora might influence the bird's need for a dietary source of the growth factor. Harrison and Coates (1964) also found a similarity in response of chicks to fish solubles and penicillin, although the growth effects of the two substances were independent and additive. In later (unpublished) work they observed no significant increase in growth of germ-free chicks given fish solubles. Pasteurized droppings from conventional birds, given by mouth at one day of age to chicks hatched germ-free, resulted in a severe depression that was partially alleviated by dietary fish solubles.

Thus although these experiments give further evidence for the existence of a growth factor in fish solubles, they indicate that it is not essential in the absence of a gut microflora.

C. GROWTH-DEPRESSING EFFECT OF RAW SOYA BEAN MEAL

Although soya beans and some other leguminous seeds are excellent sources of protein for animal feeding, they support very poor growth accompanied by hypertrophy of the pancreas, unless they are first subjected to a heat treatment. In the raw state they are known to contain toxic factors, including trypsin inhibitors, that are destroyed during the heating process. Their detrimental effect on growth cannot be completely explained in terms of inhibition of trypsin and consequent impairment of protein digestion.

Reports that antibiotic supplementation of the diet partially overcame the growth depression by raw soya in rats, chicks and poults (Braham et al., 1959; Linerode et al., 1961) gave rise to the suggestion that the gut flora may be concerned in its harmful effects. Comparison of the effects of raw soya on germ-free and conventional chicks have strengthened this view (Miller and Coates, 1966). Germ-free chicks grew only slightly less well on diets containing raw soya bean meal than on similar diets with heated meal; growth of conventional birds given the raw meal was significantly less than that of their germ-free hatchmates or controls given the heated meal. The raw soya meal diets caused the same degree of pancreatic enlargement in both environments. This effect, which is probably a response to the trypsin-inhibitors in the raw meal, is therefore independent of microbial action but the growth depression is much exacerbated by the presence of a microflora. However, it has been suggested (Jayne-Williams and Coates, 1968) that the two effects may not be entirely separate, since the inhibition of tryptic digestion could lead to the presence in the lower gut of undigested protein which, in the conventional bird, might offer a suitable substrate for organisms that have a detrimental effect on growth. Impaired fat absorption in chicks given raw soya bean meal (Nesheim et al., 1962) suggests some similarity with the "malabsorption syndrome" (Eyssen and De Somer, 1963) associated with the establishment of a growth-depressing component in the gut flora.

V. Miscellaneous Metabolic Reactions

A. OROTIC ACID

Orotic acid is a precursor in the biosynthesis of pyrimidines in animal tissues, but excessive amounts in the diet of rats induce fatty livers accompanied by alterations in the hepatic lipids and concentration of nucleotides. The condition can be reversed by the addition of small quantities of adenine to the diet (Handschumacher et al., 1960; Windmueller, 1965). Studies in germ-free and conventional rats given diets with 1% orotic acid have been made to determine whether the intestinal microflora is concerned in the metabolism of orotic acid and its action on the liver (Windmueller *et al.*, 1965). In the absence of a microflora rats absorbed 66% of the dietary orotic acid and excreted the remainder in the faeces. In conventional rats the level of orotic acid in the faeces was initially low and fell to zero after three to five days on the diet. Inoculation of the germ-free rats with a faecal extract from the conventional rats resulted in the disappearance within three days of orotic acid from their faeces. Thus the ability of microorganisms to catabolize orotic acid in the alimentary tract is clearly established. However, it is apparently not related to the development of the fatty liver condition, which occurred to the same extent in germ-free and conventional rats and was similarly prevented by administration of adenine.

B. CHOLINE

The role of microbial enzymes in the formation of urinary trimethylamine from choline has been investigated by Prentiss *et al.* (1961). When choline chloride was given intragastrically to conventional rats, the equivalent of 60% of the administered dose was excreted in the urine as trimethylamine or its oxide. Corresponding germ-free rats excreted little or no trimethylamine. In further tests, orally-administered trimethylamine was quantitatively excreted by both groups, hence the failure to detect trimethylamine in the urine of germ-free rats given choline could not be accounted for by its conversion to other substances or by differences in excretion.

From these experiments it can be concluded that the degradation of dietary choline to trimethylamine is brought about by microbial action in the gastrointestinal tract. They were undertaken as part of an investigation into the problem of liver cirrhosis that develops in rats maintained for long periods on a choline-free diet. Since the cirrhotic condition can be prevented or delayed by addition of antibiotics to the diet, there is reason to suppose that components of the gut flora may be concerned in its establishment. It is somewhat surprising, therefore, that germ-free rats have been shown to develop liver cirrhosis more rapidly than their conventional controls (Levenson *et al.*, 1960). Supplementation of the cirrhogenic diet with choline prevented cirrhosis in both germ-free and conventional animals (Levenson and Tennant, 1963). It might have been expected that in conventional rats, possessing a flora capable of degrading choline, cirrhosis would occur earlier. In discussing the apparently paradoxical results, these authors postulate the synthesis of an anticirrhotic factor by intestinal bacteria.

C. FORMATE OXIDATION

Baggiolini *et al.* (1964) followed the oxidation of ^{14}C-formate in germ-free rats and again in the same animals five to seven days after administration of a suspension of faeces from conventional rats. On each occasion a single dose of formate was given by stomach-tube and the amount of $^{14}CO_2$ in the expired air was taken as a measure of formate oxidation. The average rate of oxidation in the steady state period, expressed as μmol $^{14}CO_2/100$ g rat/hr, was 112·5 in the germ-free animals and 136·0 after contamination. The higher rate in the contaminated animals was almost certainly attributable to the oxidative activity of the gut microorganisms. The rate of oxidation of ^{14}C-formate by samples of caecal contents from the contaminated animals, expressed in the same terms, was of the right order to support this conclusion. Earlier work with conventional rats (Aebi *et al.*, 1957) had demonstrated the existence of a hepatic and an extra-hepatic component in the *in vivo* oxidation of formate. It seems likely that the extrahepatic activity is largely contributed by the intestinal bacteria.

D. BILE PIGMENTS

The concept that the reduction of bilirubin to urobilins is effected entirely by microbial action in the intestine has been substantiated by Gustafsson and Lanke (1960) who detected no urobilin in the faeces or urine of germ-free rats. Furthermore, the amount of bilirubin excreted in the faeces of germ-free rats was equal to the sum of the bilirubin and urobilins excreted by the conventional rats. Appearance of urobilins in the faeces followed association of germ-free rats with suspensions of faeces from conventional animals. Attempts to identify organisms responsible for the conversion met with partial success. A *Clostridium*-like organism, so far unidentified, when introduced into germ-free animals caused them to excrete urobilins. Its effect was increased when a strain of *E. coli* was simultaneously introduced. However, even in animals infected with the two organisms, urobilin production was only about a third as much as in conventional controls.

VI. Conclusions

Investigations with gnotobiotes have indicated a number of metabolic processes in which microorganisms of the alimentary tract play a major role, but some important aspects of nutrition have received little study in the germ-free animal. Virtually nothing is known, for instance, about the digestion of carbohydrates or the utilization of protein in the absence of a gut flora.

The effects on the host of the metabolic activities of its microflora may be beneficial or the reverse. For example, the high incidence of urinary calculi in germ-free rats indicates that the gut flora has a regulatory effect on calcium balance. Conversely, experiments with young chicks suggest that the conventional flora may hinder development of the bird's full growth potential.

Microbial synthesis of vitamins undoubtedly occurs in the conventional animal, although it is doubtful whether significant amounts of the products are available to the host without coprophagy. Less direct effects of the flora on nutrient economy are evidenced by the decreased need for ascorbic acid by the guinea pig and increased requirement of iron by the rabbit in the germ-free state.

The germ-free animal is, by nature, a well-defined entity and its basic nutritional and metabolic processes should, in due course, be clearly established. The conventional animal is less definable and its characteristics may vary according to diet and conditions of management. Experiments with antibiotics, unidentified growth factors and raw soya bean meal suggest that the effects of incorporating non-nutrient additives into the diet of conventional animals may be brought about through modification of the "normal" intestinal flora.

REFERENCES

Aebi, H., Frei, E., Knab, R. and Siegenthaler, P. (1957). *Helv. physiol. pharmac. Acta*, **15**, 150–167.

Baggiolini, M., Aebi, H., Sacquet, E. and Charlier, H. (1964). *Helv. physiol. pharmac. Acta*, **22**, 53–65.

Barnett, B. D. and Bird, H. R. (1956). *Poult. Sci.* **35**, 705–710.

Braham, J. E., Bird, H. R. and Baumann, C. A. (1959). *J. Nutr.* **67**, 149–158.

Brambel, C. D. (1960). *Int. Congr. Nutr. 5, Washington, D.C. Abstracts*, p. 25.

Coates, M. E., Fuller, R., Harrison, G. F., Lev, M. and Suffolk, S. F. (1963a). *Br. J. Nutr.* **17**, 141–150.

Coates, M. E., Gregory, M. E., Porter, J. W. G. and Williams, A. P. (1963b). *Proc. Nutr. Soc.* **22**, xxvii.

Coates, M. E., Ford, J. E. and Harrison, G. F. (1968). *Br. J. Nutr.* (in press).

Combe, E. and Sacquet, E. (1966). *C. r. hebd. Séanc. Acad. Sci., Paris*, Sér. D. **262**, 685–688.

Combe, E., Penot, E., Charlier, H. and Sacquet, E. (1965). *Annls Biol. anim. Biochim. Biophys.* **5**, 189–206.

Daft, F. S., McDaniel, E. G., Harman, L. G., Romine, M. K. and Hegner, J. R. (1963). *Fedn Proc. Fedn Am. Socs exp. Biol.* **22**, 129–133.

Dahlquist, A., Bull, B. and Gustafsson, B. E. (1965). *Archs Biochem. Biophys.* **109**, 150–158.

Desplaces, A., Zagury, D. and Sacquet, E. (1965). *C. r. hebd. Séanc. Acad. Sci., Paris* **260**, 4821–4824.

Ducluzeau, R., Raibaud, P., Dickinson, A. B., Sacquet, E. and Mocquot, G. (1966). *C. r. hebd. Séanc. Acad. Sci., Paris*, Sér. D. **262**, 944–947.

Edwards, H. M. and Boyd, F. M. (1963). *Poult. Sci.* **42**, 1030.

Evrard, E., Hoet, P. P., Eyssen, H., Charlier, H. and Sacquet, E. (1964). *Br. J. exp. Path.* **45**, 409–414.

Eyssen, H. and De Somer, P. (1963). *J. exp. Med.* **117**, 127–138.

Eyssen, H. and De Somer, P. (1965). *Ernährungsforschung* **10**, 264–273.

Eyssen, H. and De Somer, P. (1967). *Poult. Sci.* **46**, 323–333.

Forbes, M. and Park, J. T. (1959). *J. Nutr.* **67**, 69–84.

Geever, E. P. and Levenson, S. M. (1964). *Experientia* **20**, 391–392.

Glas, J. E. and Gustafsson, B. E. (1963). *Acta radiol.* **1**, 363–368.

Gustafsson, B. E. (1959). *Ann. N.Y. Acad. Sci.* **78**, 166–174.

Gustafsson, B. E. and Lanke, L. S. (1960). *J. exp. Med.* **112**, 975–981.

Gustafsson, B. E. and Norman, A. (1962). *J. exp. Med.* **116**, 273–284.

Gustafsson, B. E., Daft, F. S., McDaniel, E. G., Smith, J. C. and Fitzgerald, R. J. (1962). *J. Nutr.* **78**, 461–468.

Handschumacher, R. E., Creasey, W. A., Jaffe, J. J., Pasternak, C. A. and Hankin, L. (1960). *Proc. natn. Acad. Sci. U.S.A.* **46**, 178–186.

Harrison, G. F. and Coates, M. E. (1964). *Br. J. Nutr.* **18**, 461–466.

Huhtanen, G. N. and Pensack, J. M. (1964). *Poult. Sci.* **43**, 1331.

Jayne-Williams, D. J. and Coates, M. E. (1968). *In* "International Encyclopedia of Food and Nutrition", Vol. 17. Pergamon Press, Oxford. (In press.)

Lepkovsky, S., Wagner, M., Furuta, F., Ozone, K. and Koike, T. (1964). *Poult. Sci.* **43**, 722–726.

Lepkovsky, S., Furuta, F., Ozone, K., Koike, T. and Wagner, M. (1966). *Br. J. Nutr.* **20**, 257–261.

Lev, M. and Forbes, M. (1959). *Br. J. Nutr.* **13**, 78–84.

Levenson, S. M. and Tennant, B. (1963). *Fedn Proc. Fedn Am. Socs exp. Biol.* **22**, 109–119.

Levenson, S. M., Crowley, L. V., Horowitz, R. and Malm, O. J. (1959). *J. biol. Chem.* **234**, 2061–2062.

Levenson, S. M., Brown, N. and Horowitz, R. (1960). *Int. Congr. Nutr. 5, Washington, D.C. Abstracts*, p. 14.

Levenson, S. M., Tennant, B., Geever, E., Laundy, R. and Daft, F. S. (1962). *Archs int. Med.* **110**, 693–702.

Lindstedt, G., Lindstedt, S. and Gustafsson, B. E. (1965). *J. exp. Med.* **121**, 201–213.

Linerode, P. A., Waibel, P. E. and Pomeroy, B. S. (1961). *J. Nutr.* **75**, 427–434.

Luckey, T. D., Pleasants, J. R., Wagner, M., Gordon, H. A. and Reyniers, J. A. (1955). *J. Nutr.* **57**, 169–182.

Miller, H. T. and Luckey, T. D. (1963). *J. Nutr.* **80**, 236–242.

Miller, W. S. (1967). *Proc. Nutr. Soc.* **26**, x.

Miller, W. S. and Coates, M. E. (1966). *Proc. Nutr. Soc.* **25**, iv.

National Institute for Research in Dairying (1966). *Rep. natn. Inst. Res. Dairy.*, p. 90.

Nesheim, M. C., Garlich, J. D. and Hopkins, D. T. (1962). *J. Nutr.* **78**, 89–94.

Phillips, B. P., Wolfe, P. A. and Gordon, H. A. (1959). *Ann. N.Y. Acad. Sci.* **78**, 183–207.

Prentiss, P. G., Rosen, H., Brown, N., Horowitz, R., Malm, O. J. and Levenson, S. M. (1961). *Archs Biochem. Biophys.* **94**, 424–429.

Reddy, B. S. and Wostmann, B. S. (1966). *Archs Biochem. Biophys.* **113**, 609–616.

Reddy, B. S., Pleasants, J. R., Zimmerman, D. R. and Wostmann, B. S. (1965a). *J. Nutr.* **87**, 189–196.

Reddy, B. S., Wostmann, B. S. and Pleasants, J. R. (1965b). *J. Nutr.* **86**, 159–168.

Valencia, R., Sacquet, E., Raibaud, P., Han, N'G. C. and Charlier, H. (1965). *C. r. hebd. Séanc. Acad. Sci., Paris* **260**, 6439–6442.

Warren, K. S. and Newton, W. L. (1959). *Am. J. Physiol.* **197**, 717–720.

Wiech, N. L. and Hamilton, J. G. (1966). *Fedn Proc. Fedn Am. Socs. exp. Biol.* **25**, 321.

Windmueller, H. G. (1965). *J. Nutr.* **85**, 221–229.

Windmueller, H. G., McDaniel, E. G. and Spaeth, A. (1965). *Archs Biochem. Biophys.* **109**, 13–19.

Wostmann, B. S. and Bruckner-Kardoss, E. (1966). *Proc. Soc. exp. Biol. Med.* **121**, 1111–1114.

Wostmann, B. S. and Knight, P. L. (1961). *J. Nutr.* **74**, 103–110.

Wostmann, B. S. and Knight, P. L. (1965). *J. Nutr.* **87**, 155–160.

Wostmann, B. S., Knight, P. L. and Reyniers, J. A. (1958). *J. Nutr.* **66**, 577–586.

Wostmann, B. S., Knight, P. L. and Kan, D. F. (1962). *Ann. N.Y. Acad. Sci.* **98**, 516–527.

Wostmann, B. S., Knight, P. L., Keeley, L. L. and Kan, D. F. (1963). *Fedn Proc. Fedn Am. Socs exp. Biol.* **22**, 120–124.

Zagury, D., Desplaces, A., Sacquet, E., Ghata, J. and Charlier, H. (1962). *Presse méd.* **70**, 929.

Chapter 9

Lipid Metabolism

THOMAS F. KELLOGG AND BERNARD S. WOSTMANN

Lobund Laboratory and Department of Microbiology,
University of Notre Dame,
Notre Dame, Indiana, U.S.A.

This review includes the literature published up to June 1967 concerning the lipid metabolism of germ-free mammalian and avian species. The subject matter covered includes fatty acids, neutral sterols and bile acids.[1] Fat-soluble vitamins are discussed elsewhere (Chapter 8).

I. Fatty Acids

Studies in this area have been prompted by the blind loop syndrome in man resulting in steatorrhoea. The observation that this condition could be alleviated by oral administration of antibiotics led to an investigation of the effect of the intestinal microflora on types and amounts of fatty acids excreted in the faeces. Evrard *et al.* (1964) reported that the faecal fatty acids of germ-free rats fed either fat-free or 6% corn oil-containing diets contained 16, 18, and 20 carbon atoms. Conventional rats on similar diets excreted relatively large amounts of fatty acids with 14, 15, and 17 carbon atoms in addition to those fatty acids excreted by germ-free animals. Additional unidentified acids supposedly were either branched-chain or cyclopropane fatty acids comparable to those identified in many

[1] For nomenclature, see Appendix, p. 194.

bacterial species (O'Leary, 1962). The latter unidentified fatty acids which represented 10–25% of the total faecal acids excreted by the conventional rat were saturated free fatty acids. In conventional rats, fed on a diet containing 9% corn oil, more than 75% of the faecal fatty acids were saturated, whereas in comparable germ-free animals only 43% were saturated (Evrard et al., 1965). Faecal fatty acid patterns of these animals were similar to those reported above. Nearly 8% of the total fatty acid excretion of the conventional rat consisted of unidentified fatty acids, probably branched-chain, hydroxy- or cyclopropane-fatty acids. These materials were totally absent in the faeces of germ-free groups.

Eyssen (1966) reported that the faecal fatty acid pattern of the germ-free rat fed fat-free diets contained only those fatty acids found in tissues, particularly myristic, palmitic, stearic, oleic and linoleic acids. A gas chromatogram of conventional rat faeces showed a complex mixture of at least 14 fatty acids. The most striking difference between the two groups was the complete absence in the fatty acid pattern of the faeces of germ-free animals of a number of peaks which eluted before palmitic and between palmitic and stearic acids. Since the patterns of these unidentified acids were not modified by hydrogenation, they apparently were saturated. Relatively more unsaturated fatty acids (18 : 1 and 18 : 2) occur in the faeces of germ-free than in the faeces of conventional rats. This observation is difficult to reconcile with the hypothesis that the saturated faecal lipids arise only from selective intestinal absorption of unsaturated fatty acids and rejection of the saturated acids. Saturated fatty acids have been reported to be less easily absorbed (Fernandes et al., 1962). As Eyssen (1966) has stated, there is no conclusive evidence of extensive conversion of oleic into stearic acid in the gut of man or laboratory animals. Although unsaturated fatty acids with 18 carbon atoms are partially hydrogenated to stearic acid in the rumen of the sheep (Reiser, 1951; Shorland et al., 1955), the organisms concerned may be peculiar to ruminants. However, microorganisms from the caecum of the sheep have also been shown to hydrogenate unsaturated fatty acids (Ward et al., 1964) hence it seems possible that the caeca of other species may similarly harbour organisms capable of effecting the transformation.

The presently available data indicate that with fat-free diets quantitative faecal fat excretion is the same in conventional and germ-free rats (Evrard et al., 1964; Evrard et al., 1965). However, Evrard et al. (1965) have reported that upon feeding of a diet containing 6% corn oil, conventional rats excrete 42% more fatty acids than their germ-free counterparts. The germ-free rats did not exhibit an increase of faecal fat output with increased dietary fat since the excretion in animals fed on either 0 or 6% corn oil was essentially the same. Wiech and Hamilton

(1966) reported that no difference could be detected in absorption of linoleic-1-^{14}C acid between germ-free and conventional rats fed on a 12% corn oil diet. Fasting serum triglycerides of germ-free rats, however, were much higher (124 mg% vs 44 mg%) than the conventional controls.

The *free* fatty acid content of the faeces of germ-free and conventional rats given either 0 or 6% corn oil in the diet indicated that the amount of lipid in the diet has no particular effect on the free fatty acid content of the faeces (Evrard et al., 1965); however, the germ-free animals excrete consistently more free fatty acids than do the conventional rats. Eyssen (1966) concluded that "... in the intact rat the qualitative alterations induced in the faecal fatty acid pattern by the microflora are more pronounced than the quantitative modification of the total faecal fat output".

The faecal fatty acids of chicks (Eyssen, 1966) were reported to reflect the composition of the dietary fat and to be relatively unaltered by the intestinal microflora. This was thought to be due to the absence of a true large bowel and the very short transit time (2–5 hr) in the chicken. The type of intestinal microflora, however, markedly influences the total faecal fat output in chickens (Eyssen and DeSomer, 1967).

The crude fat content of both hard and soft faeces of germ-free rabbits is significantly lower than that of conventional rabbits. This is reflected in a significantly higher coefficient of apparent digestibility in germ-free than in conventional rabbits (Yoshida et al., 1968).

For discussion of the effects of the blind loop syndrome in rats fed various diets in either the conventional or gnotobiotic state, see the papers of Evrard et al. (1965) and Hoet et al. (1963). An extensive discussion of bacterial lipids has been published by Asselineau (1962).

II. Neutral Sterols

A. SERUM AND LIVER CHOLESTEROL LEVELS

No agreement exists with regard to the serum and liver cholesterol levels of comparable germ-free and conventional mammals and birds. Table I gives the values reported by various workers. Danielsson and Gustafsson (1959) reported a significantly higher serum cholesterol level in rats than in corresponding conventional animals. Eyssen (1966) also reported higher serum levels in germ-free than in conventional rats; however, the number of animals used was small and no statistical evaluation of the data was reported.

Wostmann and Wiech (1961) studied the serum and liver cholesterol levels of rats as a function of age on a semi-synthetic diet. They reported that rats aged 30–35 days had a significantly higher serum cholesterol level in the germ-free group than in the conventional group. These data

TABLE I

Serum and liver cholesterol levels of comparable germ-free (g-f) and conventional (conv.) groups of rats and chicks

Reference	Serum Cholesterol mg% g-f	Serum Cholesterol mg% conv.	Liver Cholesterol mg/100 g g-f	Liver Cholesterol mg/100 g conv.	Species	Age
Danielsson and Gustafsson (1959)	110 (15)[a]	77 (9) s	—	—	rat	?
Dymsza et al. (1965)	88 (5)	78 (5) s[c]	241	221[c]	rat	42 days
Eyssen (1966)	122 (3)	97 (3)	240	282	rat	?
Wostmann and Weich (1961)[b]	156 (8)	98 (10) s	367	309 s	rat	30–35 days
Wostmann and Weich (1961)	106 (19)	128 (21) s	403	326 s	rat	90–110 days
Wostmann and Weich (1961)	107 (14)	127 (13) s	—	—	rat	90–110 days
Wostmann and Weich (1961)	153 (4)	145 (8)	532 (4)	410 (10) s	rat	15–18 mo.
Wostmann and Kan (1964)	104 (10)	125 (10) s	—	—	rat	90–110 days
Wostmann et al. (1966)	110 (10)	119 (10)	394	336 s	rat	90–110 days
Edwards and Boyd (1963)	218 (12)	179 (12)	—	—	chick	14 days
Coates et al. (1965)	155 (24)	96 (24) s	—	—	chick	28 days
Coates et al. (1965)	181 (24)	171 (24)	—	—	chick	28 days
Coates et al. (1965)	151 (16)	119 (24) s	—	—	chick	28 days
Coates et al. (1965)	171 (22)	163 (22)	—	—	chick	28 days
Wostmann (unpublished)	192 (8)	148 (10) s	—	—	chick	35 days
Wostmann (unpublished)	136 (8)	152 (19)	—	—	chick	70 days

[a] Number of animals in ( ).   [b] See Table II for corrected data.
[c] Conventional animals were polyassociated ex-germ-free rats. See Dymsza et al., 1965.
"s" Indicates values significantly different from those of conventional controls.

TABLE II

Total cholesterol levels in germ-free and conventional Lobund rats of various ages. Diet L-356. All males except where indicated. Data in mg% ±S.D.M. Number of animals in parenthesis

Age in days	Serum		Liver		Aorta	
	Germ-free	Conventional	Germ-free	Conventional	Germ-free	Conventional
25–30	152 ±10 (6)	153 ±7 (13)	367 ±12 (8)[b]	309 ±14 (8)[b]	178 ±7 (15)	186 ±3 (5)
100	106 ±4 (10)[a]	123 ±5 (12)[a]	403 ±18 (8)[c]	326 ±7 (17)[c]	183 ±7 (9)	181 ±9 (8)
200	149 ±7 (10)	143 ±6 (9)				
500	153 ±9 (4)	145 ±6 (7)	532 ±50 (4)[d]	410 ±30 (10)[d]	201 ±5[e] 5 male 6 female	242 ±22[e] 4 male 4 female

[a-e] Groups with common letter superfix show statistically significant difference.

were later discovered to be incorrect, as the dietary history of the mothers of these animals had been different (Wostmann, unpublished observations). The corrected data are given in Table II. The only differences in serum cholesterol levels between germ-free and conventional males occurs around 100 days of age when the conventional group has a higher level (see also Wostmann and Kan, 1964 and Wostmann et al., 1966). Coates et al. (1965) reported studies conducted with four week-old chickens. In all four experimental series, the germ-free chicks had a higher serum cholesterol level than their conventional counterparts, although this difference was significant in only two of these series. Edwards and Boyd (1963) had found similar results, but Wostmann (unpublished observations) reported the levels to vary with age (see Table I).

Most investigators agree that germ-free liver cholesterol concentrations are higher than concentrations found in conventional animals. Dymsza et al. (1965) have reported the isolation of several species of microorganisms which as associates in gnotobiotic rats raised serum cholesterol levels. The liver cholesterol levels were not significantly altered. Wostmann (unpublished observations) has shown a slowly increasing cholesterol concentration in the aorta of both germ-free and conventional rats (Table II) with conventional rats reaching a significantly higher level by 500 days of age. This may reflect accumulated physical damage to the aorta as a result of occasional penetration of microorganisms into the circulation of the conventional animals.

Additional studies with gnotobiotic animals should lead to a clearer understanding of the role of the intestinal microflora in altering serum and liver cholesterol levels. Regardless of disagreement about serum levels, three different groups were in basic agreement on one factor. Eyssen et al. (1967), Coates et al. (1965) and Wostmann et al. (1966) all agreed that the total of serum plus liver cholesterol pools was essentially the same in the germ-free and the conventional animal. Thus, the major difference seen between the germ-free and the conventional animals was a shifting of pools and the major point left to be reconciled is which direction the pool shift takes.

B. METABOLISM OF CHOLESTEROL

Wostmann et al. (1966) have reported a study of differences in metabolism and elimination of cholesterol between germ-free and conventional rats. These studies were prompted by reports (see below) which indicated that in germ-free and in antibiotic-treated conventional rats, the rate of faecal excretion of bile acids was greatly decreased. Bile acids in the faeces of germ-free rats are conjugated and basically unmodified from the biliary bile acids. The half life of taurocholic acid in the germ-free rat was five times longer than in the conventional animal (Gustafsson *et al.*, 1957). This difference suggested greater reabsorption of the unmodified bile acids leading to a larger bile acid pool (Gustafsson *et al.*, 1960). Beher *et al.* (1961) had proposed that a feedback mechanism controlled by the size of the bile acid pool would affect cholesterol metabolism. Thus, with increased bile acid pool sizes due to increased reabsorption in the germ-free animal, it is to be expected that less cholesterol would be converted to bile acid than in the corresponding conventional animal.

Wostmann *et al.* (1966) demonstrated that over 75% of the ^{14}C in the 26 position of cholesterol, which had been oxidatively removed from the cholesterol molecule in the intact rat, appeared as expired $^{14}CO_2$. [26–^{14}C] Cholesterol was injected intravenously into germ-free and

TABLE III

Percentage distribution of ^{14}C after intravenous administration of [26 – ^{14}C] cholesterol to germ-free and conventional male rats

	Germ-free	Conventional	P
Expired air	19·0 ±1·2	29·7 ±2·0	0·01
Carcass 3β-OH sterol	48·9 ±1·7	42·5 ±1·7	0·02
Other	3·7 ±1·1	4·1 ±1·2	N.S.
Liver: 3β-OH sterol	10·0 ±0·6	7·5 ±0·3	0·01
Other	1·7 ±0·4	2·2 ±0·5	N.S.
Faecal Extract	2·6 ±0·4	4·6 ±0·4	0·01
Residual (urine, caecal contents, blood, etc.)	3·2 ±0·1	2·3 ±0·1	0·01
Total recovery of original dose	89·1 ±1·8	92·9 ±2·0	N.S.

Experimental period: 72 hr. Diet: Semisynthetic L-356. 10 animals per group. Mean values ±SEM. (Wostmann *et al.*, 1966).

conventional rats. The animals were placed in metabolism cages and expired CO_2 was collected for 72 hr. After the collection period the animals were sacrificed and the distribution and form of the isotope within the body were determined. The results are presented in Table III.

In the presence of a normal intestinal microflora, the rate of oxidative conversion of the isotope from cholesterol to expired CO_2 was at least 50% greater than that found in the germ-free animal (30% of the administered dose *vs* 19%, respectively). This more rapid rate of catabolism was verified by the fact that the ^{14}C labelled 3-β-hydroxy-sterol content of the carcass and liver of the conventional rat was lower than that of its germ-free counterpart. The faecal excretion of the isotope, which represents cholesterol and its neutral sterol metabolites, was nearly twice as high in conventional as in germ-free rats. These studies indicate that cholesterol is more rapidly metabolized to bile acid in the presence of a "normal" intestinal microflora. This is in agreement with other reports (see below) which state that the elimination of bile acid and cholesterol is greater in conventional than in germ-free animals.

C. FAECAL EXCRETION OF NEUTRAL STEROLS

Bondzynsky (1896) isolated a sterol from the feces of the dog which had the chemical properties of a dihydro-cholesterol. Subsequent work by others showed that this material was not the same dihydro-cholesterol as was produced by catalytic hydrogenation of cholesterol *in vitro*. Bondzynski speculated that this material arose as a result of bacterial hydrogenation of cholesterol during its passage through the intestinal tract and gave it the name coprosterin (coprostanol). Coleman *et al.* (1956) showed that the neutral sterols in the faeces of conventional rats are a mixture of sterols, coprostanol being quantitatively the most important. Using suspensions of faecal microorganisms, Snog-Kjaer *et al.* (1956) and Coleman and Baumann (1957) demonstrated that cholesterol could be reduced to coprostanol *in vitro*.

Danielsson and Gustafsson (1959) confirmed the bacterial origin of coprostanol by demonstrating that this material was not present in the faeces of germ-free rats. In this last study, germ-free rats fed a steroid-free diet excreted about 2·6 mg/day of neutral sterol in the faeces. The absence of coprostanol in the faecal excretion of germ-free animals was verified by Evrard *et al.* (1964). Evrard *et al.* (1965) investigated the faecal excretion of the neutral sterols of germ-free and conventional rats fed a corn oil-containing diet and concluded that in addition to cholesterol the germ-free rats excreted unmodified dietary sterols. Comparable conventional rats excreted the neutral sterols, cholesterol, β-sitosterol and campesterol and the copro- or 5-β-analogues of these compounds. In addition 3-keto derivatives were also excreted. Kellogg and Wostmann (1967a) confirmed the absence of coprostanol and of the coprostanone series of compounds in the faecal excretion of germ-free rats. The endogenous neutral sterol excretion of germ-free rats on a semipurified diet averaged nearly 13 mg/day/kg body weight (Kellogg and Wostmann, unpublished data). This appears to agree well with Danielsson and Gustafsson (1959) (see above). Gustafsson *et al.* (1966a) have reinvestigated the intestinal and faecal sterols of germ-free and conventional rats. Their conclusions support their earlier reports. They identified cholesterol, lathosterol, methostenol, campesterol, stigmasterol and β-sitosterol as the predominant sterols in the faeces of germ-free rats. The latter three sterols were contributed by diet. The same six sterols were found in the small intestine and the caecum, cholesterol being by far the major sterol in the small intestine. No coprostanol or coprostanol analogues of the plant sterols were found in the germ-free rats.

Kellogg (1965) reported that the *in vitro* fermentation of cholesterol to coprostanol by intestinal microorganisms was inactive in the presence of carbohydrate and speculated that carbohydrate can block, by some unknown mechanism, the bacterial reduction of cholesterol to coprostanol. This observation was confirmed *in vivo* by Kellogg and Wostmann (1966) who demonstrated a direct relationship between the residual carbohydrate in the faeces of the animal and the conversion of cholesterol to coprostanol.

III. Bile Acids

In 1955 Lindstedt and Norman reported that after oral administration of ^{14}C labelled cholic acid to conventional rats, the faeces contained not only this labelled acid but in addition a number of other labelled bile acid derivatives. Little of the labelled material was excreted in the conjugated form. Feeding of chemotherapeutic substances (Norman 1955) changed this pattern. Less total bile acid was excreted and almost all of the label was excreted in the form of taurocholic acid. These data indicated a profound influence of the intestinal microflora on bile acid metabolism.

Gustafsson *et al.* (1957) studied the turnover and nature of faecal bile acids in germ-free, conventional and monoassociated rats. The diet consisted of casein, wheat starch, 10% arachis oil and supplemental vitamins and minerals. One to two mg of the sodium salt of $[24-^{14}C]$ cholic acid were autoclaved in water solution and administered *per os*. The animals were kept in metabolism cages and faeces were collected over 24 hr periods. The bile acids were analysed for radioactivity after separation on a partition chromatography column. In the three germ-free animals studied, all of the isotope was found in a single fraction with elution characteristics of taurocholic acid. No other radioactive bile acid was indicated.

Monoassociation of one of the germ-free rats with *Aspergillus niger* caused no changes in the faecal bile acid excretion pattern. However, dual association with *A. niger* and *Clostridium perfringens (welchii)* type E resulted in the excretion of large quantities of free cholic acid indicating that these organisms *in vivo* were capable of splitting the bile acid conjugate bond. Faecal excretion of the labelled cholic acid by the germ-free rats was appreciably slower than the excretion of labelled cholic acid and its bacterial derivatives by their conventional littermates (T-$\frac{1}{2}$[1], respect-

[1] T-$\frac{1}{2}$ is defined as the length of time required for one half of the isotope to be excreted.

ively, 11 and 2 days) and compared well with that reported for conventional animals given chemotherapeutic substances in the diet. Association with cither *A. niger* and/or *C. perfringens (welchii)* caused no marked difference in the rate of excretion of labelled cholic acid.

Gustafsson *et al.* (1960) reported on the faecal bile acid excretion after oral administration of ^{14}C-labelled cholic acid in conventional and germ-free rats and in rats monoassociated with *Escherichia coli*. The mean daily excretion was 5·1 mg for the conventional, 1·9 mg for the germ-free and 2·2 mg for the monoassociated group. They identified 3α, 12α-dihydroxy-7-keto-cholanic acid and 3α, 7β, 12α-trihydroxy cholanic acid as metabolites of cholic acid formed in the monoassociated rats. The 7-keto acid formed by microbial enzymes was slowly reduced by the liver to 3α, 7β, 12α-trihydroxy cholanic acid. The authors reported an isotope excretion half life (T-$\frac{1}{2}$) of two days in the conventional group, six to seven days in the germ-free group and five to six days in the monoassociated group. An extensive tabulation is given of the distribution, biological half life and daily excretion of labelled bile acids following the administration of [24–14C] cholic acid. Later Gustafsson and Norman (1962) reported on the physical form of bile acids in the gastrointestinal tract of germ-free and conventional rats. Two days after administration of [24–14C] cholic acid to germ-free and conventional rats, the intestinal contents were separated by centrifugation at $25,000 \times g$. The bile acids throughout the intestinal tract of the germ-free rats were recovered in the supernatant. In conventional rats almost all labelled bile acids in the small intestine were recovered in the supernatant, whereas 27 to 44% of the labelled bile acids in the caeca were found in the sediment. The labelled sedimentary bile acids from the caeca had a higher per cent of deoxycholic and 3α-hydroxy-12-keto-cholanic acid than those present in the supernatant.

Kellogg and Wostmann (1967b) reported that over 90% of the faecal bile acids of the germ-free rat were composed of taurocholic acid and tauro-β-muricholic acid. Trace amounts of other unidentified materials were present, one of which recently has been identified as tauro-α-muricholic acid. Studies in progress have indicated that the faecal bile acid excretion of the germ-free male Wistar rat, 90–120 days old, maintained on a semisynthetic diet, was approximately 11 mg/kg body wt/day. Our studies, as well as those of others (Grundy *et al.*, 1965), have shown a conventional faecal bile acid excretion of nearly 20 mg/kg body wt/day. The biliary bile acids of the germ-free rat consisted of chenodeoxycholic acid, α-muricholic acid, β-muricholic acid and cholic acid, all as taurine conjugates. Trace quantities of other materials were found by gas chromatography but were not identified. Cholic and β-muricholic acids have been shown to be systemic end products of bile acid metabolism whereas chenodeoxycholic and α-muricholic acids may be further metabolized by the liver (Thomas *et al.*, 1964). The observation that only systemic end products are excreted in quantity in the faeces of germ-free rats implies a generally higher level of reabsorption and a more complete metabolism of bile acids in the germ-free than in the conventional rat.

In a number of earlier papers it had been stated that cholic acid and chenodeoxycholic acid would be the only systemic metabolites to be observed in the faecal excretion of animals in the absence of a bacterial microflora. However, studies over the last ten years have resulted in the isolation of a number of 3, 6, 7, trihydroxy bile acids from the bile of conventional rats and the study of Kellogg and Wostmann cited above indicates that α-muricholic and β-muricholic acids are present in the germ-free biliary and faecal excretion. Deoxycholic acid has been generally considered to arise from the bacterial 7-dehydroxylation of cholic acid and lithocholic acid a corresponding product from chenodeoxycholic acid. Recent reports, however, (Mitropoulos and Myant, 1967) indicate that rat liver homogenates can synthesize lithocholic acid from cholesterol. These authors also reported the synthesis *in vitro* of 3α, 6β-dihydroxy-5β-cholanic acid and on the basis of these studies proposed a new scheme of metabolic pathways for the 3, 6, 7 trihydroxy bile acids in the rat. Since 3α, 6β-dihydroxy-5β-cholanic acid would be expected to arise from the bacterial 7-dehydroxylation of the muricholic acids and lithocholic can arise from a similar dehydroxylation of chenodeoxycholic acid, the germ-free animal would be an excellent tool with which to investigate the metabolism of these substances.

An extensive investigation has been published by Gustafsson *et al.* (1966b) on the *in vitro* microbiological 7α-dehydroxylation of bile acids. Using chenodeoxycholic acid as a substrate, he was able to show that isolated microorganisms were capable of removing the 7α-hydroxyl group as well as oxidizing the hydroxyl groups at carbons 3 and 7 to keto groups. A recent report by Kallner (1967) indicates that intestinal microorganisms are capable of converting deoxycholic acid to allodeoxycholic acid. Since the deoxycholic acid arises from microbiological modification of cholic acid and allodeoxycholic acid is formed from deoxycholic, it appears that a series of microbiological reactions is responsible for the formation of allodeoxycholic acid from cholic acid. Allodeoxycholic acid is one of the major constituents of gall stones.

Since the faecal excretion of bile acids in conventional animals is considerably greater than the excretion in both conventional animals fed chemotherapeutic agents and germ-free animals, bacterial action must be responsible for this increase. Which of the many changes caused by the intestinal microorganisms are responsible for this increase in excretion is not known at this time.

IV. Studies on Lipid Metabolism Utilizing Germ-free Animals

Studies by Kritchevsky *et al.* (1959) revealed that the germ-free chick fed on either a glucose or starch diet containing 3% cholesterol had a higher serum cholesterol level than its conventional control. When given sucrose as a carbohydrate, both germ-free and conventional groups showed comparable serum levels that were higher than the concentration found in germ-free birds fed on either glucose or starch. This could indicate a "germ-free" type of metabolism in chicks fed sucrose but the high values found in the germ-free sucrose groups indicated a systemic effect on serum cholesterol levels as well. Coates *et al.* (1965) also studied the effect of carbohydrate on serum cholesterol levels in chicks. Their conclusions are similar to those of Kritchevsky *et al.* (1959). However, they also found a significantly higher serum cholesterol level in germ-free chicks fed on sucrose than in conventional sucrose-fed controls. In addition they investigated the effect of diets high in butter fat and maize oil. They concluded that in the presence of dietary cholesterol, the hypercholesteraemic effect of butter fat was highly significant but that there was no difference between germ-free and conventional birds. Thus, the major effect of butter fat on exogenous cholesterol was caused by a mechanism in which the gut flora was not involved.

Edwards and Boyd (1963) investigated the rise in serum cholesterol levels of chicks given lithocholic acid in the diet. Previous work had raised the possibility that the action of lithocholic acid was mediated through some effect of the intestinal microflora. When the experiments were repeated utilizing germ-free chicks, there was a marked increase in serum cholesterol levels in both germ-free and conventional groups. Thus, the hypercholesteraemic activity of lithocholic acid was not dependent on the intestinal microflora. Wostmann and Kan (1964) reported that conventional rats exhibited 20–30% lower serum cholesterol levels when fed on a commercial laboratory chow diet as compared to a semisynthetic diet. Utilizing germ-free rats to assay the activity of the material, they showed that the activity resided in the hexane-extractable portion of the crude diet. Since this activity could be demonstrated in germ-free animals, the unidentified factor apparently had a direct systemic effect on the metabolism of the animal.

Eyssen *et al.* (1966b) reported that 0·2% neomycin administered to conventional chicks fed on a diet containing 0·2% cholesterol reduced serum cholesterol levels by nearly 20%. N-methylated neomycin (a neomycin derivative without antibiotic properties) was also an effective cholesterol lowering agent, while neutral N-acetylated neomycin and streptomycin with three basic groups in the molecule were both inactive. The condensation product of streptomycin that bears seven basic groups was, however, a very effective compound. On the basis of these results, the authors concluded that the effect of these substances on serum cholesterol was independent of their antimicrobial activity but appeared to be related to the polybasic nature of the molecule. Observations that neomycin and N-methylated neomycin precipitate solutions of bile acids *in vitro* neutralize the toxic effect of lithocholic acid *in vivo* and increase bile acid excretion led to the conclusion that the serum cholesterol lowering effect of these substances was due to their ability to form bile salts with bile acids in the gut.

These conclusions were verified by Eyssen *et al.* (1966). Feeding of neomycin to germ-free chicks for two weeks reduced serum and liver cholesterol levels 25 to 50% and caused a two-fold increase in the faecal bile acid output; feeding N-methylated neomycin had similar effects.

V. Concluding Remarks

Although generalizations often lead to more confusion than clarification, certain repetitive themes are constant throughout the work reported here. It is clear for instance that the presence of an intestinal microflora results in an increase in types and amounts of excreted lipids as compared to the patterns found in the germ-free animal. Thus, a major effect of the intestinal microflora on these lipids seems to be one of forming derivatives of the original parent material although there is excellent evidence that in regard to the bacterial fatty acids some materials are synthesized *de novo*. The increased excretion of all types of lipids by the conventional animals compared to their germ-free counterparts indicates that some of the lipids thus formed are less absorbable than their substrates.

Use of the germ-free animal as a tool to investigate the effects of the intestinal microflora, either by comparison with the conventional animal or through selective association of the otherwise germ-free animal, is becoming more widespread. In unpublished studies from our laboratory on the faecal excretion of bile acids and neutral sterols in germ-free, conventional and monoassociated animals, we observed that the standard deviations in the germ-free animal series were several times smaller than those for the conventional animals. Thus, in the study of systemic metabolism where components of the intestinal microflora can be intimately involved, one obtains greater sensitivity in comparisons between groups if germ-free animals or gnotobiotes are utilized rather than conventional animals. This would be of particular importance where the materials under study are either expensive or difficult to isolate or purify.

We can expect an increasing number of experiments involving systemic cholesterol and bile acid metabolism to be conducted in germ-free animals. Intrinsic microflora patterns, as well as the response of the flora to variations in the diet, vary widely from one laboratory to another. Utilization of animals devoid of an intestinal microflora or harbouring only defined, selected microbial species would make results between different laboratories much more comparable. Indeed, the utilization of germ-free animals seems to be required in circumstances where the systemic metabolism of materials which may also be synthesized or modified by the intestinal microflora is to be studied.

VI. APPENDIX

The bile acids and sterols mentioned in this chapter are listed below with the recognized systematic names in parentheses.

3α, 12α-dihydroxy-7-ketocholanic acid (3α, 12α-dihydroxy-7-oxo-5β-cholanoic acid)

3α, 7β, 12α-trihydroxycholanic acid (3α, 7β, 12α-trihydroxy-5β-cholanoic acid)

lithocholic acid (3α-hydroxy-5β-cholanoic acid)

chenodeoxycholic acid (3α, 7α-dihydroxy-5β-cholanoic acid)

cholic acid (3α, 7α, 12α-trihydroxy-5β-cholanoic acid)

α-muricholic acid (3α, 6β, 7α-trihydroxy-5β-cholanoic acid)

β-muricholic acid (3α, 6β, 7β-trihydroxy-5β-cholanoic acid)

deoxycholic acid (3α, 12α-dihydroxy-5β-cholanoic acid)

3α-hydroxy-12-ketocholanic acid (3α-hydroxy-12-oxo-5β-cholanoic acid)

3α, 6β-dihydroxy-5β-cholanic acid (3α, 6β-dihydroxy-5β-cholanoic acid)

allodeoxycholic acid (3α, 12α-dihydroxy-5α-cholanoic acid)

lathosterol (5α-cholest-7-en-3β-ol)

methostenol (4α-methyl-5α-cholest-7-en-3β-ol)

β-sitosterol (24α-ethylcholest-5-en-3β-ol)

campesterol (24α-methylcholest-5-en-3β-ol)

stigmasterol (24α-ethylcholest-5, 22-dien-3β-ol)

REFERENCES

Asselineau, J. (1962). "Les Lipides Bactériens", Hermann, Paris.

Beher, W. T., Baker, G. D., Anthony, W. L. and Beher, M. E. (1961). *Henry Ford Hosp. med. Bull.* **9**, 201–213.

Bondzynsky, S. (1896). *Chem. Ber.* **29**, 476.

Coates, M. E., Harrison, G. F. and Moore, J. H. (1965). *Ernährungsforschung* **10**, 251–256.

Coleman, D. L. and Baumann, C. A. (1957). *Archs Biochem. Biophys.* **72**, 219–225.

Coleman, D. L., Wells, W. W. and Baumann, C. A. (1956). *Archs Biochem. Biophys.* **60**, 412–418.

Danielsson, H. and Gustafsson, B. (1959). *Archs Biochem. Biophys.* **83**, 482–485.

Dymsza, H. A., Stoewsand, G. S., Enright, J. J., Trexler, P. C. and Gall, L. C. (1965). *Nature, Lond.* **208**, 1236–1237.

Edwards, H. M. and Boyd, F. M. (1963). *Proc. Soc. exp. Biol. Med.* **113**, 294–295.

Evrard, E., Hoet, P. P., Eyssen, H., Charlier, H. and Sacquet, E. (1964). *Br. J. exp. Path.* **45**, 409–414.

Evrard, E., Sacquet, E., Raibaud, P., Charlier, H., Dickinson, A., Eyssen, H. and Hoet, P. P. (1965). *Ernährungsforschung* **10**, 257–263.

Eyssen, H. (1966). *Int. Congr. Microbiol. IX*, Symposia, 329–340.

Eyssen, H. and DeSomer, P. (1967). *Poultry Sci.* **46**, 323–333.

Eyssen, H., Evrard, E., and Van den Bosch, J. (1966a). *Life Sciences* **5**, 1729–1734.

Eyssen, H., Evrard, E. and Vanderhaeghe, H. (1966b). *J. Lab. clin. Med.* **68**, 753–768.

Eyssen, H., Sacquet, E., Evrard, E. and Van den Bosch, J. (1967). *Life Sciences* (In press.)

Fernandes, J., Van de Kamer, J. H. and Weijers, H. A. (1962). *J. clin. Invest.* **41**, 488–494.

Grundy, S. M., Ahrens, E. H., Jr. and Miettinen, T. A. (1965). *J. Lipid Res.* **6**, 397–410.

Gustafsson, B. E. and Norman, A. (1962). *Proc. Soc. exp. Biol. Med.* **110**, 387–389.

Gustafsson, B. E., Bergstrom, S., Lindstedt, S. and Norman, A. (1957). *Proc. Soc. exp. Biol. Med.* **94**, 467–471.

Gustafsson, B. E., Norman, A. and Sjövall, J. (1960). *Archs Biochem. Biophys.* **91**, 93–100.

Gustafsson, B. E., Gustafsson, J. A. and Sjövall, J. (1966a). *Acta chem. scand.* **20**, 1827–1835.

Gustafsson, B. E., Midtvedt, T. and Norman, A. (1966b). *J. exp. Med.* **123**, 413–432.

Hoet, P. P., Joossens, J. V., Evrard, E., Eyssen, H. and DeSomer, P. (1963). *In* "Biochemical Problems of Lipids" Proc. 7th Int. Conf., Birmingham, 1962 (A. C. Frazer, ed.). pp. 73–83. Elsevier Pub. Co., Amsterdam.

Kallner, A. (1967). *Acta chem. scand.* **21**, 315–321.

Kellogg, T. F. (1965). Unpublished Ph.D. thesis, University of Wisconsin Library, Madison, Wisconsin, U.S.A.

Kellogg, T. F. and Wostmann, B. S. (1966). *Biochim. biophys. Acta* **125**, 617–619.

Kellogg, T. F. and Wostmann, B. S. (1967a). *Fedn Proc. Fedn Am. Socs exp. Biol.* **26**, 653.

Kellogg, T. F. and Wostmann, B. S. (1967b). *Proc. Indiana Acad. Sci.* **76**. (In press.)

Kritchevsky, D., Koman, R. R., Guttmacher, R. M. and Forbes, M. (1959). *Archs Biochem. Biophys.* **85**, 444–451.

Lindstedt, S. and Norman, A. (1955). *Acta physiol. scand.* **34**, 1–10.

Mitropoulos, K. A. and Myant, N. B. (1967). *Biochem. J.* **103**, 472–479.

Norman, A. (1955). *Acta physiol. scand.* **33**, 99–107.

O'Leary, W. M. (1962). *Bact. Rev.* **26**, 421–447.

Reiser, R. (1951). *Fedn Proc. Fedn Am. Socs exp. Biol.* **10**, 236.

Shorland, F. B., Weenink, R. O. and Johns, A. T. (1955). *Nature, Lond.* **175**, 1129–1130.

Snog-Kjaer, A., Prange, I. and Dam, H. (1956). *J. gen. Microbiol.* **14**, 256–260.
Thomas, P. J., Hsia, S. L., Matschiner, J. T., Doisy, E. A., Jr., Elliott, W. H.,
    Thayer, S. A. and Doisy, E. A. (1964). *J. biol. Chem.* **239**, 102–105.
Ward, P. F. V., Scott, T. W. and Dawson, R. M. C. (1964). *Biochem. J.* **92**, 60–68.
Wiech, N. L. and Hamilton, J. G. (1966). *Fedn Proc. Fedn Am. Socs exp. Biol.*
    25, 321.
Wostmann, B. S. and Kan, D. F. (1964). *J. Nutr.* **84**, 277–282.
Wostmann, B. S. and Wiech, N. L. (1961). *Am. J. Physiol.* **201**, 1027–1029.
Wostmann, B. S., Wiech, N. L. and Kung, E. (1966). *J. Lipid Res.* **7**, 77–82.
Yoshida, T., Pleasants, J. R., Reddy, B. S. and Wostmann, B. S. (1968). *Br. J.*
    Nutr. (In press.)

Chapter 10

Defence Mechanisms in Germ-free Animals

Part I. Humoral Defence Mechanisms

B. S. WOSTMANN

Lobund Laboratory, Department of Microbiology,
University of Notre Dame, Notre Dame, Indiana, U.S.A.

I. Introduction

In the germ-free animal we have achieved a major simplification of the most important experimental system used in the Life Sciences: the intact animal. The removal of the viable microflora has resulted in the elimination of many of the variables inherently associated with an ever changing microbial population. Therefore the germ-free animal appears to be especially suited to study the genetically determined potentialities of the organism and its reaction to various exogenous conditions, including the introduction of microbial entities.

The germ-free animal, by our present day standards, is actually an animal free of bacteria and bacteria-like organisms (PPLO), protozoa, fungi and metazoan parasites. Thus far leukaemogenic virus has been demonstrated in all strains of germ-free mice but not in germ-free rats (Pollard, 1965a). Diets for germ-free animals frequently contain ill-defined materials. They may be heavily contaminated with microorganisms before steam or radiation sterilization and are often antigenic *per se* (Sell, 1964b). Sterilization procedures may produce macro-molecular condensation products of unknown properties. Free contact with animals of the same species resulting in ingestion of excreta and other body products is seldom prevented. Except in very few cases, this germ-free animal has been exposed to the usual exogenous stimuli other than those specifically originating from a viable microflora. As a result a number of authors have reported limited reticuloendotheliar stimulation and specific antibody activity in germ-free rats, mice and chickens, especially in older animals (Springer *et al.*, 1959; Wagner, 1959; Cohen *et al.*, 1963).

TABLE I

Mesenteric lymph node of the mouse: mg tissue/100 g body weight and ratio between absolute amounts of various types of cells in non-stimulated germ-free and conventional mice. Diet L-462 (Wostmann, 1959)

	mg/100 g body wt	Cell type[a]	Ratio Conv/Gf
Germ-free	$42 \cdot 2 \pm 3 \cdot 3$	Blast	8·4
Conventional	134 ± 11	Plasma blast	12·3
		Large lymphocyte	12·3
		Immature plasma cell	14·4
		Mature plasma cell	8·3
		Medium lymphocyte	3·7
		Small lymphocyte	3·0

[a] See Olson and Wostmann (1966a).

TABLE II

Proliferation time (hr) of lymphoid cells in non-stimulated and stimulated germ-free and conventional mice using tritiated thymidine index method

Stimulation	Conventional		Germ-free	
	Mesenteric	Cervical	Mesenteric	Cervical
Control	21·9	21·0	21·9	nd[a]
7S HGG	18·6	16·5	20·4	nd
S. typhimurium (formalinized)	13·2	8	12·3	nd

[a] Not determined.
Data from Table I, Olson and Wostmann (1966a).

The response of the reticuloendothelial system (RES) to a germ-free environment varies from species to species, but all show a reduction in lymphocytic tissue (Gordon, 1959; Gordon *et al.*, 1966), fewer reaction zones (Thorbecke *et al.*, 1957; Pollard, 1965b) and per cent less plasmacytic elements (Thorbecke *et al.*, 1957; Olson and Wostmann, 1966a) (Table I). *In vivo* labeling with tritiated thymidine indicates that only part of the nodal sinuses permit free entrance of the label (Olson and Wostmann, 1966a). Generation time of lymphoid cells, however, appears similar in germ-free and conventional mice and is shortened in a comparable way by antigenic stimulation (Table II) (Olson and Wostmann, 1966b).

II. Serum Proteins

Assuming that immune proteins originate only upon antigenic stimulation, the serum protein pattern of the germ-free animal seems to reflect the above mentioned observations. Gamma globulins are almost always low in concentration but never seem to be entirely absent

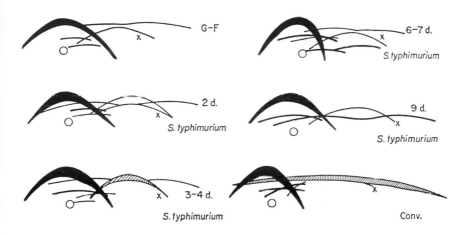

Fig. 1 Immunoelectrophoretic pattern of gamma globulin range of germ-free and conventional rats, and of rats monoassociated with S. typhimurium. Site of application and transferrin arc indicated. Lines reaching transferrin arc from α globulin range omitted. Developed with anti-rat rabbit antiserum (see text).

(Gustafsson and Laurell, 1958; Wostmann, 1961; Sell, 1964b; Asovsky, unpublished). Some difference of opinion appears to exist about the γ_M concentration in mouse serum, as one study shows "normal" levels (Arnason et al., 1964), while another claims this fraction to be below detectable limits (Fahey and Sell, 1965). Differences in strain, diet and husbandry may have contributed to these results. The germ-free rat demonstrates a serum gamma globulin pattern which appears remarkably independent of the dietary regime. Immunoelectrophoresis shows that all fractions present in the gamma range of the conventional animal are represented but at a much lower concentration. An exception may be the γ_X protein (Fig. 1), presumably of non-immune origin (Wostmann, unpublished). In the case of the germ-free mouse (Asovsky, unpublished), guinea pig (Sell, 1964b; Wostmann, 1961) and rabbit (Wostmann, 1961), the serum protein pattern is obviously affected by dietary materials. In germ-free rabbits and guinea pigs the gamma globulin concentration is substantially increased by incorporation of bovine milk proteins into the diet (Wostmann, 1961). In guinea pigs the 7S γ_1 fraction appears to be especially diet dependent (Sell, 1964b). A similar observation has been made in germ-free mice (Asovsky, unpublished).

TABLE III

Serum proteins of germ-free rats fed a water-soluble, "antigen-free" diet compared with values obtained from rats fed a solid, practical-type diet. Age: approximately 2–5 months. Technique: cellulose acetate paper electrophoresis; values in mg/100 ml serum. S.D.M. values given

Status	Diet	Alb.	1	2	3	4[a]	5	Total
Germ-free (6♂ ♀)	Water sol.	2693 ±77	1530 ±122		479 ±43	876 ±56	95 ±11	5680 ±150
Germ-free (17 ♂)	L-462	2349 ±72	1274 ±54	539 ±26	489 ±40	708 ±27	123 ±11	5490 ±130
Conv. (14 ♂)	L-462	2421 ±73	1026 ±55	488 ±26	512 ±29	970 ±35	819 ±58	6220 ±110

[a] Contains transferrin.
(Data from Table I, Wostmann et al. (1965).

At the Lobund Laboratory we have succeeded in raising germ-free rats from birth to adulthood on water-soluble, chemically defined, filter-sterilized diets. Surprisingly, with the formulations used thus far, these animals did not demonstrate much lower gamma globulin levels than those found in animals maintained on practical type diets like L-462 (Table III) (Wostmann et al., 1965). The immunoelectrophoretic pattern of the serum of these rats again showed all immune proteins found in the serum of the conventional animal (Fig. 1) (Wostmann, unpublished). Although at the time of the above mentioned experiments the chemical purity of the lipid components of these diets was not yet optimal and therefore a certain antigenicity from that source not excluded, the most recent studies with diets of more precise chemical definition essentially show the same result. These data suggest that at least in the rat a certain low rate of production of the various "immunoglobulins" may occur even in the absence of exogenous antigenic stimulation.

The fractional rate of catabolism of gamma globulin of germ-free chickens and of guinea pigs appears comparable to that found in the conventional animals (Sell, 1964a; Wostmann and Olson, 1964b). The half lives of 7–8 days (guinea pigs) and 4·2 days (chickens) represent mostly the turnover of the 7S gamma globulins. Detailed studies in mice have revealed a dependence of 7S gamma globulin turnover on concentration of the 7S class of gamma globulins (Fahey and Sell, 1965). Besides differences between 7S γ_1, 7S γ_{2A} and 7S γ_{2B}, the data indicate much longer half lives for these fractions in the germ-free animal. No such effect was indicated for the rapidly metabolizing γ_A and γ_M fractions.

Upon exposure to a balanced microflora characteristic for the species, the ex-germ-free rat shows an early increase in α globulins which at that moment presumably reflects the formation of an Acute Phase Protein (Wostmann and Gordon, 1960). β Globulins increase shortly thereafter but for at least 2 weeks the gamma component of the electrophoretic pattern determined by free boundary gel or paper techniques remains at the low germ-free level of 100 or 150 mg/100 ml. This indicates that at least the 7S γ_2 component of the rat is slow to react to the stimulation of the balanced microflora and does not reach its "conventional" level until after 4 to 6 weeks (Wostmann and Gordon, 1960; Gustafsson and Laurell, 1959).

Monoassociation of the germ-free rat with a number of well-tolerated microbial species hardly changes the serum protein pattern, although the serum demonstrates specific agglutination titres comparable to those found in conventional rats in which these species occur (Wostmann, 1961; Gustafsson and Laurell, 1959; Wagner and Wostmann, 1961). Almost invariably the monoassociated rat harbours a microbial population comparable in number with that of the conventional animal. In the germ-free chicken, on the other hand, monoassociation with Streptococcus faecalis results in gamma globulin levels comparable to those found in conventional birds. Agglutinin titres, however, were of the same order as those found in rats monoassociated with the same microorganism (Table IV) (Wagner and Wostmann, 1961). This presumably indicates that in the rat antibody formation is highly specific and therefore very effective. Limited data on monoassociated rabbits seem to place this animal in the same class as the chicken (Wostmann, unpublished).

After association with a virulent strain of Salmonella typhimurium (Lobund # 750A) the electrophoretic pattern of the ex-germ-free rat shows very specific changes. Within 48 hr albumin drops sharply but α globulins increase, presumably because of the formation of Acute Phase

Protein. The typical gamma range of the cellulose acetate electrophoresis pattern, representing mostly 7S γ_2 globulin, appears to increase only after day 6, a few days after anti-*S. typhimurium* agglutinins in the serum become obvious. Thereafter the gamma fraction steadily increases to reach "conventional" levels at 6 to 8 weeks (Wostmann, unpublished).

TABLE IV

Serum globulin fractions[a] and antibacterial agglutinins in rats and chickens reared germ-free, conventionally or monoassociated. Number of animals in parentheses

	Serum globulin fractions mg/100 ml			Agglutinins	
	Alpha	Beta	Gamma	Reciprocal titer	Test organism
Rats (age approx. 4 mo)					
Germ-free	58 ±5	299 ±14	126 ±9 (16)	0 (16) 0 (16)	*Str. faecalis* *L. casei*
Str. faecalis monoassociated	86 ±6	353 ±26	147 ±13 (5)	32–256 (6)	*Str. faecalis*
L. casei monoassociated	85	239	152 (2)	32–128 (3)	*L. casei*
Conventional	188 ±25	591 ±34	523 ±52 (12)	2–256 (44)	*Str. faecalis* *L. casei*
Chickens (age approx. 2½ mo)				0 (13)	
Germ-free	95 ±11	362 ±20	319 ±29 (7)	0 (7)	*Str. faecalis*
Str. faecalis monoassociated	89 ±11	414 ±49	770 ±113(8)	32–128 (5)	*Str. faecalis*
Conventional	122 ±9	490 ±27	701 ±48 (12)	8–32 (9)	*Str. faecalis*

[a] Antweiler free boundary microelectrophoresis technique.
Data from Tables I and II. Wagner and Wostmann (1961).

As mentioned earlier the germ-free rat, even when raised on water-soluble, chemically defined diets, appears to contain low but demonstrable amounts of all immune globulins found in the conventional animal. The same would appear true for the germ-free mouse and the germ-free guinea pig (reared on solid diet), although depending on dietary and other conditions one or another fraction may fall below the normal limit of detection (Sell, 1964b; Fahey and Sell, 1965). However, the patterns of the germ-free serum of all three species reveal a well defined arc in the lower gamma globulin range which does not seem to coincide with any of the recognized immune proteins (Fahey and Sell, 1965; Wostmann and Olson, 1964a; Wostmann, unpublished). In the pattern of the conventional animals this arc will often fuse with the heavy 7S γ line and appear as "spurs" emerging from that line. Evidence in the guinea pig (Wostmann and Olson, 1964a) and especially in the rat point to a non-immune nature of this slowly migrating gamma globulin. The fact that upon association of the germ-free rat with *S. typhimurium* this arc becomes quite prominent at day 3 to 4, only to be reduced to its former appearance at the end of the first week, indicates a protein fraction which greatly increases during the first days of the heavy stress of pathogenic intrusion (Fig. 1) (Wostmann, unpublished). These proteins may be similar to a "stress" protein migrating in the slow gamma range which is formed upon administration of milligram quantities of *E. coli* endotoxin to rats (Lawford and White, 1964).

III. Antibody Formation

The consensus of opinion appears to be that the present germ-free animal responds adequately to exogenous antigenic stimulation. The graft *vs* host reaction does not seem any different from that seen in the conventional animal (Salomon, 1965). Monoassociation with microbial species results in the formation of homologous agglutinins (Table IV), although in the case of the "milder" associates their appearance may be delayed until the second week of association. In older animals of most of the germ-free species tested, low levels of antimicrobial agglutinins are found which indicate exposure to dead forms in diet and bedding material (Wagner, 1959; Springer *et al.*, 1959; Cohen *et al.*, 1963).

Germ-free rats monoassociated with *Lactobacillus casei* by oral inoculation at weaning developed homologous circulating and salivary antibody (Wagner and Wostmann, 1961; Wagner and Orland, 1964). However, agglutinating antibody was not demonstrated in their progeny, suggesting possible acquired immunological tolerance to *L. casei* in animals born to parents harbouring this organism. Germ-free rats foster-suckled at less than one day of age on *L. casei*-associated lactating dams developed antibody, suggesting a pre-natal rather than a neo-natal effect underlying this "tolerant state" (Wagner, 1966). Simulated rather than classical acquired immunological tolerance seemed to be involved since progeny born to *L. casei*-associated parents could be stimulated to produce antibody by parenteral injection of the antigen. The simulated tolerant state was not observed in similar experiments with gnotobiotic rats monoassociated with *Str. faecalis*.

The reduced size of the RES may, however, limit the magnitude of the primary response. Both the germ-free mouse and the germ-free chicken (Fig. 2) show a primary response to protein and bacterial antigen which at peak titre is one half or less of that seen in the conventional animal (Wostmann and Olson, 1964b; Horowitz *et al.*, 1964; Olson and Wostmann, 1966b).

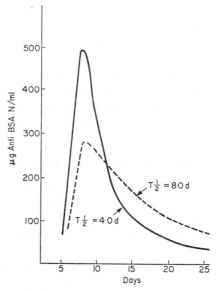

FIG. 2 Precipitating antibody in germ-free and conventional chickens following a single stimulation with 40 mg BSA/kg body weight. $T\frac{1}{2}$ indicates half value time of precipitating antibody (see text). Conventional ———, germ-free ----.

Even while we conclude, mostly from data obtained with germ-free mice and chickens, that the antibody response of the germ-free animal is essentially adequate, it must be pointed out that the pattern of the response appears to be influenced by the absence of microbial stimulation. Studies at the Lobund Laboratory have revealed that the primary anti-bovine serum albumin (BSA) precipitin formation in the germ-free chicken not only demonstrates a lower peak response but also a much longer half value time[1] after peak response (Fig. 2). In conjunction with the fact that the half lives of the gamma globulins of germ-free and

[1] Time needed for antibody concentration to drop to 50% of an earlier value.

conventional chickens are comparable (approximately 4 days, see p. 201), the latter observation leads to the conclusion that in the germ-free bird, in the absence of stimuli originating from a viable microflora, anti-BSA precipitin formation continues beyond its usual end point, the time approximately when peak titres are reached (Wostmann and Olson, 1964b). Administration of a second unrelated protein antigen, human

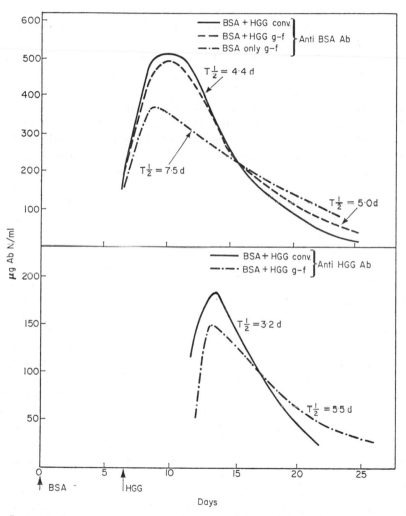

FIG. 3 Antibody response to a single dose of BSA (40 mg/kg body weight) followed 7 days later by the administration of a single dose of HGG (40 mg/kg body weight). $T\frac{1}{2}$ indicates half value time of antibody concentration after peak titre (see also Table V).

gamma globulin (HGG), some time after the first (BSA) led to a suppression of the prolonged anti-BSA response in the germ-free chicken, as reflected by its reduced half-value time (Fig. 3). However, the resulting anti-HGG response again showed the prolongation characteristic for the germ-free state. It appears that in the absence of a major influx of antigenic material of microbial origin the antibody-forming system will be committed much longer to the specific response to a certain antigen, unless a later influx of antigenic material of a different nature crowds out the response to the first antigen (Table V) (Wostmann and Olson, 1965b).

Intracellular digestion of bacterial material by the peritoneal macrophage of the germ-free mouse appears slower than in the conventional animal (Bauer et al., 1964). Although the role of the macrophage in the transfer of antigenic information is by no means decided (Fishman and Adler, 1961; Franzl and Morello, 1966) the above phenomenon could conceivably affect the antibody response of the germ-free animal. From the available data on antibody response in germ-free animals, such an effect is not obvious.

All results mentioned thus far reinforce the impression of adequacy of the immune response of the germ-free animal. Almost all of these data have been obtained from germ-free animals that were still exposed to smaller or greater doses of exogenous, antigenic material. In most cases these germ-free rats, mice and rabbits have shown a low but definite and specific bactericidal action of the serum (Landy and Weidanz, 1964; Ikari, 1964). These (and other) antibodies seem to be absent only from the germ-free colostrum-free piglet reared on a non-antigenic diet (Sterzl, 1966). The presence of "natural antibody" supposedly points to a certain amount of exposure to somatic antigen of Gram-negative microorganisms or to materials inducing a similar response (Michael et al., 1962). This in turn must have led to some stimulation of the RES. With our present potential of maintaining germ-free rats directly from birth to maturity on antigen-free, chemically defined water-soluble diets the possibility of reaching adulthood with an absolute minimum in antigenic and synergistic stimulation seems attainable. None of these animals so far has been reared in absolute solitary confinement but the experiments are in progress. Preparations are also made for maintaining highly inbred strains of mice on these non-antigenic diets. These studies will answer two major questions: (a) will these animals still demonstrate immunoglobulins and, if so, which ones; (b) will they, in the absolute absence of synergistic factors like endotoxins, still react adequately to stimulation with, for example, a pure preparation of BSA.

Besides specific antibody a number of other factors aid in anti-microbial

TABLE V

Antibody response of germ-free and conventional 3–4 month-old chickens stimulated with BSA, or HGG, or with BSA followed by HGG

| Status | Number of animals | Antigen | Maximum antibody Concentration | | | | Half-value time (days)[a] | |
| | | | Anti-BSA | | Anti-HGG | | Anti-BSA | Anti-HGG |
			Day	μg N/ml	Day	μg N/ml		
G-F series I	10	BSA	8	290 ±29[b,c]			8.0[c]	
series II	5	BSA	8	387 ±58			7.5 ±0.6	
Conv. series I	10	BSA	8	500 ±74[c]			4.0[c]	
series II	5	BSA	8	668 ±117			4.1 ±0.1	
G-F series II	3	HGG			7	192 ±48		5.0 ±0.6
Conv. series II	8	HGG			7	216 ±41		3.4 ±0.2
G-F series II	10	BSA + HGG[d]	8–9	481 ±45	7	146 ±19	5.0 ±0.1	5.5 ±0.2
Conv. series II	9	BSA ± HGG[d]	8–9	509 ±78	7	170 ±28	4.4 ±0.2	3.2 ±0.1

[a] Time needed for antibody concentration to be reduced by 50%.
[b] Mean value ± standard error.
[c] Comparable data from earlier work.
[d] HGG administered 7 days after BSA.

defence. Complement in germ-free guinea pigs attains levels similar to those seen in the conventional animal (Newton *et al.*, 1960). Properdin concentrations, on the other hand, were found to be low in germ-free rats and upon association with a complete microflora, to increase slowly like the gamma globulin levels (Gustafsson and Laurell, 1960). A recent report indicates the α globulin fraction in the serum of rats and guinea pigs as promoting *in vitro* phagocytosis of *Staphylococcus aureus*. No difference in activity was found between germ-free and conventional serum (Downey and Pisano, 1966).

Although germ-free animals in general seem to be able to harbour the same viruses as their conventional counterparts, little is known about their antiviral defences. Recent studies indicate, however, that upon viral challenge germ-free mice and rats demonstrate interferon titres comparable to, or possibly even slightly higher than, those of conventional animals (DeSomer and Billiau, 1966; Fitzgerald, 1966).

IV. Conclusion

Depending on dietary and environmental conditions, germ-free animals possess an antibody production system that is quantitatively smaller, deficient in potential antibody forming cells and possibly only partly (directly) accessible. Upon antigenic stimulation this system functions well, although influence of more prolonged bacterial stimulation may be necessary to overcome the effects of the original deficiency in size and cellularity. Absence of the stimulation by a viable microflora reduces exogenous stimuli to a level which appears to prolong the effects of stimulation by a specific antigen.

REFERENCES

Arnason, B. G., Salomon, J. C. and Grabar, P. (1964). *C. r. hebd. Séanc. Acad. Sci., Paris* **259**, 4882–4885.
Bauer, H., Horowitz, R. E., Watkins, K. C. and Popper, H. (1964). *J. Am. med. Ass.* **187**, 715–718.
Cohen, J. O., Newton, W. L., Cherry, W. B. and Updyke, E. L. (1963). *J. Immun.* **90**, 358–367.
DeSomer, P. and Billiau, A. (1966). *Arch. ges. Virusforsch.* **19**, 143–154.
Downey, R. J. and Pisano, J. C. (1966). *Life Sciences* **5**, 1325–1333.
Fahey, J. L. and Sell, S. (1965). *J. exp. Med.* **122**, 41–58.
Fishman, M. and Adler, F. L. (1961). *J. exp. Med.* **114**, 837–856.
Fitzgerald, G. R. (1966). "A Comparison of the Interferon-producing Potential of Germfree and Conventional Mice and the Effect of Irradiation on Interferon Production." Doctoral Thesis. Department of Microbiology, University of Notre Dame, Indiana, U.S.A.
Franzl, R. E. and Morello, J. A. (1966). *J. Reticuloendothelial Soc.* **3**, 351–352.

Gordon, H. A. (1959). *Ann. N.Y. Acad. Sci.* **78**, 208–220.

Gordon, H. A., Bruckner-Kardoss, E., Staley, T. E., Wagner, M. and Wostmann, B. S. (1966). *Acta anat.* **64**, 301–323.

Gustafsson, B. E. and Laurell, C. B. (1958). *J. exp. Med.* **108**, 251–258.

Gustafsson, B. E. and Laurell, C. B. (1959). *J. exp. Med.* **110**, 675–684.

Gustafsson, B. E. and Laurell, A. (1960). *Proc. Soc. exp. Biol. Med.* **105**, 598–600.

Harowitz, R. E., Bauer, H., Paronetto, F., Abrams, G. D., Watkins, K. C. and Popper, H. (1964). *Am. J. Path.* **44**, 747–761.

Ikari, N. S. (1964). *Nature, Lond.* **202**, 879–881.

Landy, M. and Weidanz, W. P. (1964). *In* "Bacterial Endotoxins", pp. 275-290, (M. Landy and W. Braun, eds) Rutgers University Press, New Brunswick, New Jersey, U.S.A.

Lawford, D. J. and White, R. G. (1964). *Nature, Lond.* **201**, 705–706.

Michael, J. G., Whitby, J. L. and Landy, M. (1962). *J. exp. Med.* **115**, 131–146.

Newton, W. L., Pennington, R. M. and Lieberman, J. (1960). *Proc. Soc. exp. Biol. Med.* **104**, 486–488.

Olson, G. B. and Wostmann, B. S. (1966a). *J. Immun.* **97**, 267–274.

Olson, G. B. and Wostmann, B. S. (1966b). *J. Immun.* **97**, 275–286.

Pollard, M. (1965a). *Natn. Cancer Inst. Monogr.* No. **20**.

Pollard, M. (1965b). *Prog. med. Virol.* **7**, 362–376.

Salomon, J. C. (1965). *C. r. hebd. Séanc. Acad. Sci., Paris* **260**, 4862–4864.

Sell, (1964a). *J. Immun.* **92**, 559–564.

Sell, S. (1964b). *J. Immun.* **93**, 122–131.

Springer, G. F., Horton, R. E. and Forbes, M. (1959). *Ann. N.Y. Acad. Sci.* **78**, 272–275.

Sterzl, J. (1966). *Int. Congr. Microbiol. IX, Moscow Symposia* p. 381.

Thorbecke, G. J., Gordon, H. A., Wostmann, B. S., Wagner, M. and Reyniers, J. A. (1957). *J. infect. Dis.* **101**, 237–251.

Wagner, M. (1959). *Ann. N.Y. Acad. Sci.* **78**, 261–271.

Wagner, M. (1966). "A Study of the Effects of Specific Immunization on Experimental Dental Caries in the Gnotobiotic Rat", Doctoral Thesis. Purdue University, West Lafayette, Indiana, U.S.A.

Wagner, M. and Orland, F. J. (1964). *Proc. Indiana Acad. Sci.* **73**, 75.

Wagner, M. and Wostmann, B. S. (1961). *Ann. N.Y. Acad. Sci.* **94**, 210–217.

Wostmann, B. S. (1959). *Ann. N.Y. Acad. Sci.* **78**, 175–182.

Wostmann, B. S. (1961). *Ann. N.Y. Acad. Sci.* **94**, 272–283.

Wostmann, B. S. and Gordon, H. A. (1960). *J. Immun.* **84**, 27–31.

Wostmann, B. S. and Olson, G. B. (1964a). *Proc. Soc. exp. Biol. Med.* **116**, 914–918.

Wostmann, B. S. and Olson, G. B. (1964b). *J. Immun.* **92**, 41.

Wostmann, B. S. and Olson, G. B. (1965). *Proc. Indiana Acad. Sci.* **74**, 120–122.

Wostmann, B. S., Olson, G. B. and Pleasants, J. R. (1965). *Nature, Lond.* **206**, 1056–1057.

Part II. Cellular Defence Mechanisms*

HEINZ BAUER

Georgetown University Schools of Medicine and Dentistry,
Washington, D.C., U.S.A.

Lymphatic tissue interacts with the microbial flora in many ways throughout the lifetime of conventional animals and provides them with an important defence system against microorganisms. Since there is no living microflora in germ-free animals, their lymph nodes, spleen and other lymphatic tissue reflect the lack of stimuli derived in conventional animals from the microbial environment. Germ-free lymphatic tissue thus reveals the morphological and functional baseline characteristics from which response to microbial challenge develops. For all these reasons the lymphatic system has been studied since the early days of research with germ-free animals. Before reviewing the reactions of lymphatic tissue, a discussion of its structure, cells and function follows.

*Supported (in part) by Contract No. DA-49-193-MD-2541 and Contract No. DA-MD-49-193-64-G129 from the United States Army Research and Development Command.

I. The Functional Anatomy of Lymphatic Tissue

A. GENERAL CHARACTERISTICS

While lymphatic tissue occurs focally in many organs such as the respiratory and gastrointestinal tract, it is most highly organized and functional in the spleen and lymph nodes throughout the body. Anatomically, the spleen can be divided into the follicles (white pulp) gathered around the central arterioles and the red pulp, and the lymph nodes into cortex and medulla (Moe, 1963). The splenic follicles (Weiss, 1964) and the lymph node cortex consist of a delicate connective tissue framework which is filled with lymphocytes and reticuloendothelial cells. The red pulp of the spleen and the medulla of the lymph nodes are composed of a system of sinuses for the transport and phagocytosis of various materials and, in the rodent spleen, for haematopoiesis.

B. THE CELLS OF LYMPHATIC TISSUE

Three types of cells populate all lymphatic tissue and although functionally related and probably interdependent (Berman, 1966), they can be distinguished morphologically (Moe, 1964).

1. LYMPHOCYTES

Lymphocytes are usually classified as small, medium and large types and account for the majority of cells in lymphatic tissue. While they are morphologically simple, their turnover and function are variable (Little *et al.*, 1962) and complex (Yoffey, 1964). There is mounting evidence that lymphocytes are concerned with antibody formation (Gowans and McGregor, 1965) but conclusive proof is still lacking. Lymphocytes circulate from lymph nodes, remain in the blood stream for short periods of time and then enter lymphatic tissue in other parts of the body (Gowans and Knight, 1964). Lymphocytes may also be involved in reactions of delayed hypersensitivity (Gowens and McGregor, 1965).

2. RETICULOENDOTHELIAL CELLS

The second group of cells in lymphatic tissue consists of reticuloendothelial cells (macrophages, histiocytes) which line the sinuses or navigate them freely as macrophages. As reticulum cells they form the structural elements of lymph nodes and of the spleen. In the sinuses of lymph nodes and of the spleen, reticuloendothelial cells function mainly as phagocytes and are rich in acid phosphatase (Barka *et al.*, 1961) and other catabolic enzymes contained in lysosomes (Weissmann, 1965). After phagocytosis, macrophages undergo changes in their enzyme composition (Cohn and Wiener, 1963) which tend to enhance their digestive capacity. In addition to the phagocytosis and intracellular disposal of exogenous as well as endogenous materials, macrophages also play an important part in immunogenesis (Thorbecke and Benacerraf, 1962; Berman, 1966). This process probably begins with phagocytosis which has been attributed by some authors to a mechanism of antigen "recognition" (Vaughan, 1965), by others to a specific location of such phagocytic cells in lymphatic tissue (Miller, 1964), or as a third alternative to the physicochemical characteristics of the antigen (Cohen *et al.*, 1966). The intracellular modification of antigen which follows phagocytosis (Campbell and Garvey, 1963) then leads to the transfer of immunogenic fragments or "information" to potential antibody forming cells (Fishman *et al.*, 1963) and this sequence of events probably governs the onset, speed and efficiency of the immune response. A crucial role has been ascribed to macrophages also in delayed hypersensitivity (Wiener *et al.*, 1965).

3. PLASMA CELLS AND THEIR PRECURSORS

These cells in lymphatic tissue are most clearly involved in immune reactions. The histological (Welsh, 1962), electron microscopic (Bessis, 1961) and immunocytochemical (Vazquez, 1961) characteristics of the plasma cell are well known but their origin is still unclear (McMillan and Engelbert, 1963). While mature plasma cells produce gamma globulin, the large pyroninophilic cells which precede their appearance are thought to be transformed lymphocytes (Roberts, 1960) or derived from another precursor cell called plasmablast (Schooley, 1961), immunoblast (Dameshek, 1963), or, by international nomenclature designation, haemocytoblast (Fagraeus, 1960). A dual origin from lymphocytes and mesenchymal cells has also been postulated (McMillan and Engelbert, 1963).

Despite these genealogical problems, there is no doubt that the incidence and function of the plasma cell and its precursors are related to immune events, particularly in lymphatic tissue. During the primary immune response, i.e. on first exposure to an antigen, the large pyroninophilic cells (Berman, 1963) appear early in the cortex of lymph nodes (Baney *et al.*, 1962) or in the follicles of the spleen (Ward *et al.*, 1959). They arise whether the antigen induces delayed hypersensitivity (Oort and Turk, 1965) or humoral antibody formation (Schooley, 1961). In the latter type of response, gamma globulin producing cells develop from these precursors and antibody appears in the serum. A judgment of immunological activity in lymphatic tissue can thus be made from the number, type, distribution and protein synthesis of these antigen-responsive cells.

Another consequence of antigenic stimulation is the appearance of reaction centres (Pernis *et al.*, 1963) which are particularly prominent during the secondary response, i.e. after re-exposure to the antigen (White, 1960). These discrete areas of rapidly proliferating cells (Oort and Turk, 1965) appear in the outer cortex of lymph nodes or the splenic follicles and consist of large pyroninophilic cells and lymphocytes (Fliedner *et al.*, 1964) surrounded by or containing gamma globulin. This led Mellors and Korngold (1963) to believe that these cells form antibody but other invesigators have attributed the presence of gamma globulin in reaction centres to circulating antibody trapped by antigen sequestered in the centres (Cohen *et al.*, 1966). Current studies indicate that reaction centres function mainly as areas of antigen localization (Swartzendruber, 1966), enabling local macrophages to ingest and "prepare" antigenic materials and to pass the products on to nearby lymphocytes or other plasma cell precursors. Reaction centres, therefore, can also be used as indicators of immunological activity in lymphatic tissue.

II. The Functional Anatomy of the Thymus

A. GENERAL CHARACTERISTICS

The thymus represents a special type of lymphatic tissue which, at least in the neonate, affects the morphology of lymphatic tissue elsewhere and its responsiveness to antigenic stimuli (Karetzky and Rudolf, 1964). In man, the thymus has its greatest functional significance in infancy and later atrophies but in certain diseases it also assumes importance in the adult (Chatten, 1964). In rodents, the thymus does not involute and retains functional significance (Sherman, 1965) as a lymphocytopoietic organ (Leblond and Sainte-Marie, 1960; Azar, 1963) and possibly as the source of a humoral factor promoting immunological competence of lymphatic tissue (Osoba, 1965). Others have failed to confirm the presence of a humoral mechanism (Hvros *et al.*, 1966). Anatomically, the cortex of the rodent thymus consists of actively proliferating lymphocytes which are believed to differ subtly from those in other organs (Miller *et al.*, 1962). They migrate from the cortex to the medulla, enter the circulating pool of lymphocytes (Sherman, 1965) and "seed" peripheral lymphatic tissue (Nossal, 1964; Murray and Woods, 1965). Since lymph nodes respond to antigenic stimulation by a rapid increase in lymphocytes as well as plasma cells (Gowans and McGregor, 1965), the level of cell production in the thymus may parallel the needs of lymphatic tissue elsewhere and thus serve as another parameter for judging overall immunological activity, at least in the rodent host.

B. THE RESULTS OF THYMECTOMY

Rodents thymectomized at birth subsequently show lymphopenia (Good *et al.*, 1962), atrophic lymphatic tissue (Waksman *et al.*, 1962), an impaired immune response (Good *et al.*, 1962) and wasting (Sherman *et al.*, 1963; Karetzky and Rudolf, 1964). These deficiencies were attributed to the athymic state but others implicated chronic infections for the underdevelopment of the animals (Azar, 1964). As will be discussed later, studies in germ-free animals have deepened the understanding of these phenomena.

III. The Cytokinetics of Lymphatic Tissue

The cytokinetics of lymphatic tissue have been studied histologically (Leblond and Sainte-Marie, 1960) and by autoradiography after the administration of ^{3}H-thymidine as a label for newly synthesized DNA. In lymph nodes and the spleen, cell production probably parallels immunological activity since most labelled cells are either in reaction centres (Yoffey *et al.*, 1961) or in their vicinity (Edwards and Klein, 1961) and are mainly lymphocytes and plasma cell precursors (Berman, 1966). The mitotic rate of macrophages is low in physiological circumstances (Schooley, 1961). In the thymus of the rodent, lymphocytopoiesis is very active in the outer cortex and less prominent in the medulla (Edwards and Klein, 1961), correlating with reports of thymic cytokinetics based on other methods.

IV. Lymphatic Tissue, Thymus and Irradiation

Lymphatic tissue is readily injured by irradiation and even non-lethal doses destroy virtually all lymphocytes in the field of irradiation (De Bruyn, 1948). The cortical lymphocytes of the thymus are similarly affected (Bloom and Bloom, 1954). Macrophages and plasma cells in lymphatic tissue (Bauer *et al.*, 1963b) as well as the epithelial cells and medullary lymphocytes of the rodent thymus (Trowell, 1961) survive even high doses of irradiation. While plasma cells continue to function (Bauer *et al.*, 1963b) and the phagocytic action of macrophages remains intact (Benacerraf *et al.*, 1959), the power to destroy intracellular microorganisms is impaired, frequently resulting in systemic infection (Wensinck, 1961). If animals survive irradiation long enough, the thymus and lymphatic tissue elsewhere recover their lymphocyte population but repopulation in the rodent may be thymus-dependent (Auerbach, 1964).

V. Lymphatic Tissue in Germ-free Animals

A. GENERAL CHARACTERISTICS

The potential usefulness of studying lymphatic tissue in germ-free animals has been recognized since the beginnings of germ-free research and extensive reviews of the findings reported by earlier workers are available (Gordon, 1960; Luckey, 1963). Briefly, germ-free lymphatic tissue is comparatively smaller in those areas which in conventional animals are in close contact with the microbial flora. Histologically, germ-free lymph nodes and spleens lack large pyroninophilic cells, plasma cells and reaction centres or show such features less frequently than their conventional counterpart. More recent studies attempted to test differentially the morphology and function of all cellular elements in lymphatic tissue, gauging the effect of the microflora by systematic comparisons between germ-free and conventional animals. The difficulties of judging these dynamic processes were partly overcome by using multiple investigative parameters ranging from semiquantitative histology, histochemistry and autoradiography to fluorescent antibody techniques and serology.

B. THE CELLS

1. LYMPHOCYTES

In earlier studies reviewed in a recent book (Luckey, 1963, pp. 360–366), the lymph nodes and spleens of germ-free animals were reported as smaller and containing fewer lymphocytes than those of conventional animals. These findings were associated with a lower lymphocyte count in the peripheral blood and less retention of circulating lymphocytes by the lamina propria of the germ-free intestine. The conventional animals used for comparison in these studies were maintained under open laboratory conditions, however, and thus probably exposed to subclinical infections and their immunological consequences resulting from the unpredictable microbial environment. When conventional controls were housed and fed like the germ-free, the differences in the size and weight of lymphatic organs became insignificant (Bauer et al., 1964). Semi-quantitative histological studies under these experimental conditions showed that when lymph nodes from the same anatomic area were compared in germ-free and conventional mice, the number, distribution and density of small lymphocytes were identical. The incidence of large and medium lymphocytes generally paralleled that of the plasma cell series.

2. MACROPHAGES

Histologically and histochemically, the macrophages in the lymph nodes and spleens of germ-free and conventional mice look alike and their incidence, distribution and their content of pigments such as iron and lipochrome vary with the anatomic location of the organ or the age of the animal but not with the microbial state of the host (Bauer *et al.*, 1963). Macrophages are thus equally numerous in lymph nodes draining the oropharynx and lower intestinal tract of germ-free and conventional mice. In both types of animal the bulk of exogenous materials enters the body through these regions but their bacterial colonization in the conventional animal does not seem to increase macrophage activity further. In conjunction with the large number of macrophages which populate the cervical and mesenteric lymph nodes, the sinuses of these nodes are more numerous and wider. Meneghelli (1961) has reported similar findings in man when mesenteric lymph nodes were compared with those of the axilla.

While a review of earlier data indicated that the phagocytic cells of germ-free animals are less efficient than their conventional counterparts (Luckey, 1963, p. 380), later studies using blood clearance as the experimental model failed to confirm this difference (Thorbecke and Benecerraf, 1959; Doll, 1962). While phagocytosis seems unaffected by "inexperience" with microorganisms, biochemical studies have shown that the pulmonary alveolar macrophages of conventional animals are richer in catabolic lysosomal enzymes than their germ-free counterparts (Heise and Myrvik, 1966). This deficiency in the germ-free host may be related to a "priming" effect of inhaled microorganisms in conventional animals since the enzyme content of macrophages from the normally sterile peritoneal cavity did not exceed that of the germ-free cells. The lower content of catabolic enzymes in the cytoplasm of germ-free macrophages may have functional significance since germ-free animals are less able to destroy virulent (Hobby *et al.*, 1966) and attenuated (Suter and Kirsanow, 1962) tubercle bacilli, other bacteria (Luckey, 1963, pp. 402–419; Fauve, 1964; Stollerman *et al.*, 1965) and viruses (Dolowy and Muldoon, 1964; Schaffer, 1963).

3. PLASMA CELLS AND THEIR PRECURSORS

The major differences between the lymph nodes and spleens of germ-free and conventional animals are related to those cells which reflect immunological activity (Bauer *et al.*, 1963). Medium and large lymphocytes probably formed by activation of small lymphocytes exposed to antigen, large pyroninophilic cells, plasma cells whether identified histologically or immunocytochemically by their gamma globulin content, and reaction centres are, therefore, all rare in germ-free lymphatic tissue. When lymph nodes draining areas of heavy bacterial colonization in conventional animals are compared with their germ-free counterpart, the lymphatic tissue of the latter shows little evidence of these morphological and functional expressions of immunological activity. Germ-free lymph nodes and spleens, however, are not devoid of these findings (Thorbecke, 1959; Abrams and Bishop, 1961) particularly in mesenteric nodes (Bauer et al., 1964). This indicates that although immunological stimuli in healthy animals derive mainly from the microbial flora, the germ-free host, as Springer (1959) has pointed out, is not antigen-free. The development of synthetic diets (Pollard, 1964b, Pleasants et al., 1964) lacking potentially antigenic materials has extended the scope of these investigations (see Chapter 4).

4. SUMMARY

Morphological and functional studies in normal germ-free and conventional animals have indicated the importance of the microbial flora in determining the cellular composition and function of lymphatic tissue. The microbial environment affects the phagocytic and immunocytic elements of lymphatic tissue in a different manner. Contact with living microorganisms does not alter the number, appearance and phagocytosis of macrophages but the level of immunological activity in lymphatic tissue is determined largely by the microflora and thus remains low in the germ-free host.

C. RESPONSE TO ANTIGENIC STIMULATION

1. PARTICULATE AND SOLUBLE ANTIGENS

The immune response of germ-free animals has been tested with a variety of soluble and particulate antigens. These include the soluble proteins ovalbumin (Lerner, 1964), bovine gamma globulin (Sell, 1965), ferritin (Bauer et al., 1966b), bovine serum albumin (Wostmann and Olson, 1964; Kim et al., 1966) and human gamma globulin (Olson and Wostmann, 1965); the bacteria Escherichia coli (Horowitz et al., 1964; Ikari, 1964), Serratia marcescens (Bauer et al., 1966b), Shigella (Sprinz et al., 1961), Salmonella typhosa (Olson and Wostmann, 1965) and Mycobacterium tuberculosis (Lerner, 1964) as well as some viruses (Dolowy and Muldoon, 1964; Kim et al., 1966; Tennant et al., 1965). Antibody formation to sheep erythrocytes (Bosma et al., 1966) and the malarial parasite Plasmodium berghei were also investigated (Martin et al., 1966). All these studies showed that germ-free animals are capable of responding to antigenic materials basically like their conventional counter-part. There were some minor morphological and functional differences, however, affecting mainly the early phases of immunogenesis. After anti-genic challenge in mice (Horowitz *et al.*, 1964; Olson and Wostmann, 1965; Bauer *et al.*, 1966) and chickens (Wostmann and Olson, 1964), the characteristic cellular manifestations of antibody formation and serum antibody appeared later and progressed more slowly in the germ-free animals. Horowitz *et al.* (1964) and Bauer *et al.* (1966) attributed this lag to delayed degradation of antigen. They followed the intracellular fate of the antigens by the fluorescent antibody technique and found differences in the speed of intracellular degradation of the antigens while the time of their arrival in the node and phagocytosis were identical in both types of animal. Clear temporal relationships could be established in the germ-free and conventional animals between the disintegration of antigen in macrophages, the appearance of large pyroninophilic cells and the onset of antibody formation. These processes all began later in the germ-free animals. On the other hand, the relatively prolonged retention of antigen in the nodes and the presumably slower release of immunogenic fragments from macrophages resulted in a more sustained and some-times greater response in the germ-free than in the conventional animals.

The lifelong exposure of conventional animals to the immunological effects of the microbial flora thus enables them to initiate immuno-genesis more rapidly. This is only a minor advantage, however, and the differences between the immune response in germ-free and conventional animals are quantitative rather than qualitative. More important is the way in which these findings have confirmed the importance of macrophages in the immune response. Since the antigens appeared at the same time and were phagocytosed equally well by the lymph node macrophages of the germ-free and conventional mice, lymphatic drainage or deficient phago-cytosis cannot explain the delayed onset of antibody formation in the germ-free animals. The physical characteristics and source of the antigen also seem irrelevant since Bauer *et al.* (1966) injected killed microorgan-isms and a sterile heterologous serum protein simultaneously into the same animal and found that both were handled alike. Previous exposure to the antigens and pre-existing serum antibody were also excluded in these studies. The observed difference in the ability of macrophages to "prepare" antigen was thus intrinsic to these cells and attributable only to the microbial state of the animals prior to challenge. This difference between the macrophages of germ-free and conventional animals has shown that macrophages are an essential first step in immunogenesis and can regulate the onset and level of antibody formation.

2. BACTERIAL ANTIGENS IN THE DIET ("NATURAL" ANTIBODIES)

Antibodies have been called "natural" when they appear without obvious antigenic stimulation in normal individuals (Michael et al., 1962). They occur in many mammalian sera including those of man and react with a number of toxins, tissue cells and bacteria. There is some controversy as to the specificity of these factors (Skarnes and Watson, 1957) but recent work has confirmed their antibody character (Cohen et al., 1963). Although there are some species differences, germ-free animals also develop "natural" antibodies (Wagner, 1959; Michael et al. 1962; Cohen et al., 1963; Ikari, 1964).

No studies are available which correlate the morphology of germ-free animals or their lymphatic tissue with the presence of "natural" antibodies. If the antigens are derived from dead bacteria in the sterilized food, resultant antibodies should be produced mainly in abdominal lymphatic tissue. This is probably correct since mesenteric lymph nodes usually show slight immunological activity even in the germ-free host (Bauer et al., 1964). This low-level immune response correlates well with the low titres of "natural" antibodies (Wagner, 1959). The germ-free animal thus responds to naturally occuring antigenic stimuli of low intensity by the production of "natural" antibodies.

3. CONVENTIONALIZATION

Luckey (1963, pp. 425–428) has reviewed the results of placing germ-free animals into a conventional environment or of contaminating their food with the faeces of conventional animals. Lymphatic tissue reacts to this burst of antigenic stimulation with enlargement of the upper and lower intestinal lymph nodes (Gustafsson, 1959; Hudson and Luckey, 1964) and a rise in serum gamma globulin (Gustafsson and Laurell, 1959). Since germ-free animals generally remain clinically well during this transition period (Gordon, 1960), the increase in lymphatic tissue mass probably represents an immune response rather than significant infection. In more detailed studies, germ-free animals that were fed caecal contents of conventional animals or were selectively associated with groups of microbial species developed an immune response in their cervical and abdominal lymph nodes (Carter et al., 1965); Carter et al., 1966). This reaction appears only when three or more types of bacteria are used, however, or when the animals are exposed to a full conventional flora. The magnitude of the immune response in germ-free animals contaminated with living organisms by the oral route thus correlates with the number of bacterial species in the inoculum.

4. AUTOIMMUNITY

In addition to naturally occuring microbial antigens, germ-free animals also react to autoimmune stimuli. This type of reactivity has been reported in regional lymph nodes responding to breakdown products of skin and subcutaneous tissue damaged by a non-infectious eczema (Bauer *et al.*, 1963) and after experimental liver injury (Grabar, 1965; Bauer *et al.*, 1966a). Sterile neoplasms induced in germ-free animals by chemical carcinogens also evoke an immune response in regional lymphatic tissue (Pollard, 1964a) and the development of experimental allergic encephalo-myelitis is the same in germ-free and conventional rats (Olson and Burnstein, 1965). Finally, "runting" has been produced when newborn germ-free mice of one strain were inoculated with spleen cells of adult animals from another strain (Salomon, 1965), permitting the exclusion of microbial factors from the pathogenesis of "runt" disease.

5. THYMUS

Luckey (1963, pp. 362–366) has reviewed the earlier information pertaining to the thymus of germ-free animals. These studies showed that the gland tends to be smaller and less cellular than in conventional animals. Recent assays of cell production in the thymus using *in vivo* labeling of newly synthesized DNA with ^{3}H – thymidine demonstrated that cell turnover is slower in the thymus of germ-free mice (Burns *et al.*, 1964). This low level of thymic lymphocytopoiesis in the germ-free host correlates well with the relative inactivity of immunocytic elements, including lymphocytes, in the peripheral lymphatic tissue of germ-free animals. The slower growth rate of the thymus observed by Wilson *et al.* (1965) in germ-free as compared to conventional mice may also be related to the low level of immunological activity prevailing in the germ-free host.

Thymectomy in neonatal rats and mice impairs immune responses and causes lymphocytopenia in the blood and lymphatic tissue (Miller, 1963). Through the development of techniques for thymectomy in newborn germ-free mice (Wilson, 1963), the infectious etiology of post-thymectomy wasting disease as suggested by Azar (1964), Salvin *et al.* (1965) and Duhig (1965), was confirmed since germ-free mice thymectomized at birth developed normally (Wilson *et al.*, 1964). Exposure of these animals to a conventional flora as late as eight months after thymectomy, however, is followed by stunted growth. This indicates that the concept of impaired defence mechanisms in neonatally thymectomized animals remains valid.

Wasting disease induced by the administration of cortisone to young mice (Schlesinger and Mark, 1963) is also less severe in germ-free animals

(Reed and Jutila, 1965). In another experimental model for the production of wasting disease, i.e. the repeated injection of sterile bacterial vaccines into neonates, germ-free mice proved similarly more resistant (Ekstedt, 1964).

6. IRRADIATION

Germ-free animals withstand irradiation better than their conventional counterpart as shown by a higher median lethal dose and prolonged survival time (McLaughlin et al., 1961; Wilson, 1963). Bauer et al. (1963b) reported that the lymph nodes, spleens and thymus of germ-free mice given 550 R of whole-body irradiation recovered their lymphocyte population as rapidly as those of the surviving conventional animals. While macrophages showed no changes after irradiation, a burst of plasma cell proliferation occurred in the lymphatic tissue of all surviving animals. A similarly increased plasma cell activity has been observed by others in conventional mice (Ghossein et al., 1963), rats (Nossal, 1959; Allegretti et al., 1962) and dogs (Wohlwill and Jetter, 1953) after similar low-lethal doses of irradiation. Despite increased plasma cell activity and, presumably, gamma globulin production, the authors observed that the serum gamma globulin level rose only in the germ-free and fell in the conventional animals and cited numerous reports of a similar decrease in serum gamma globulin after low-lethal irradiation in conventional animals of several mammalian species. In the light of studies with germ-free animals, the post-irradiation plasmocytosis of lymphatic tissue cannot be attributed to the actions of the microbial flora and may result from stimuli derived from irradiation. In contrast, the loss of gamma globulin which has been attributed to irradiation (Nossal, 1959) may be caused primarily by the microbial flora, perhaps in conjunction with radiation injury.

VI. Inflammation and Healing in Germ-free Animals

Relatively few studies have utilized germ-free animals for the study of inflammation. The mobilization of polymorphonuclear leukocytes during the early phases of the inflammatory response has been investigated by Abrams and Bishop (1965) who found that sterile injury elicited an earlier and greater exudation of these cells in conventional than in germ-free mice. The delayed and diminished appearance of polymorphonuclear leukocytes at the site of injury in the germ-free animals seemed to result from local rather than systemic factors since the number of leukocytes in the blood was the same in both types of animal. This indicates a conditioning effect of the microbial flora on the vascular aspects of acute inflammation.

Rovin *et al.* (1965, 1966) observed that intraoral wounds healed equally well in germ-free and conventional mice and similar results have been reported by Miyakawa *et al.* (1958) studying skin wounds in guinea pigs. When infection occurred in their conventional guinea pigs, however, epidermal regeneration was retarded as compared to wounds which remained clean. In another experimental model, Carter *et al.* (1966) found that epidermal regeneration was retarded in skin incisions of conventional rats subjected to bilateral nephrectomy in comparison to their germ-free counterparts while the inflammatory response was more marked in the conventional animals. Inflammation and healing are thus affected differently by the microbial flora which augments the polymorphonuclear response but hampers repair.

VII. Conclusions

Bacteria are the main source of antigenic materials in healthy animals and the level of immunological activity in lymphatic tissue parallels the intensity of its exposure to the microflora.

Existence in the germ-free state does not alter macrophage morphology, distribution, phagocytic capacity or the delivery of antigenic inocula to the regional lymph nodes. While these functions are thus equally developed in germ-free and conventional mice, an enhanced capacity of macrophages to process antigens seems to result from the continuous exposure of conventional animals to the immunological effects of the microbial flora. The lack of substantial antigenic stimulation in germ-free animals fails to develop this macrophage function beyond the basic ability to degrade foreign substances. This deficiency results in a relatively delayed onset of the immune response in germ-free lymphatic tissue but the slower antigen digestion in macrophages and the presumably slower release of immunogenic fragments of antigen or "information" result in a more sustained response than in conventional animals. This modifying effect of the microbial flora on the function of lymphatic tissue during the immune response seems to be independent of previous experience with, or the nature of, the antigen. Fundamentally, however, germ-free lymphatic tissue responds to exogenous and endogenous antigens like that of conventional animals.

The lymphocytes of germ-free and conventional mice are equally susceptible to radiation injury. During the recovery period, the spleen and lymph nodes of both types of animal develop morphological evidence of an immune response. Microorganisms thus are not the main cause of this post-irradiation phenomenon. In germ-free mice, this increased immunological activity causes a rise in the serum gamma globulin level while that of conventional animals decreases despite morphological evidence of increased production. Microorganisms thus appear responsible for the loss of gamma globulin in animals surviving irradiation.

Animals and their microbial environment affect each other in many ways. Studies of lymphatic tissue in germ-free animals have served to clarify some aspects of this important symbiotic and sometimes synergistic relationship.

VIII. Acknowledgements

The author's studies cited in this chapter resulted from a collaborative effort between the Department of Germ-free Research, Division of Surgery, Walter Reed Army Institute of Research, Washington, D. C. and the Departments of Pathology, The Mount Sinai School of Medicine, New York and Georgetown University Schools of Medicine and Dentistry, Washington, D. C. The co-operation of the scientific and technical personnel of these institutions is gratefully acknowledged.

REFERENCES

Abrams, G. D. and Bishop, J. E. (1961). *Univ. Mich. med. Bull.* **27**, 136–147.
Abrams, G. D. and Bishop, J. E. (1965). *Archs Path.* **79**, 213–217.
Allegretti, N., Vitale, B. and Dekaris, D. (1962). *Int. J. Radiat. Biol.* **4**, 363–370.
Auerbach, R. (1964). *In* "The Thymus in Immunobiology" (R. A. Good and A. E. Gabrielsen, eds), pp. 107–109. Harper and Row, New York.
Azar, H. A. (1963). *Archs Path.* **76**, 653–658.
Azar, H. A. (1964). *Proc. Soc. exp. Biol. Med.* **116**, 817–823.
Baney, R. N., Vazquez, J. and Dixon, F. J. (1962). *Proc. Soc. exp. Biol. Med.* **109**, 1–4.
Barka, T., Schaffner, F. and Popper, H. (1961). *Lab. Invest.* **10**, 590–607.
Bauer, H., Horowitz, R. E., Levenson, S. M. and Popper, H. (1963a). *Am. J. Path.* **42**, 471–483.
Bauer, H., Horowitz, R. E., Paronetto, F., Einheber, A., Abrams, G. D. and Popper, H. (1963b). *Lab. Invest.* **13**, 381–388.
Bauer, H., Horowitz, R. E., Watkins, K. C. and Popper, H. (1964). *J. Am. med. Assoc.* **187**, 715–718.
Bauer, H., Paronetto, F., Porro, R. F. and Einheber, A. (1966a). *Fedn Proc. Fedn Am. Socs exp. Biol.* **25**, 667.
Bauer, H., Paronetto, F., Burns, W. A. and Einheber, A. (1966b). *J. exp. Med.* **123**, 1013–1024.
Benacerraf, N., Kivey-Rosenberg, E., Sebestyen, M. M. and Zweifach, B. W. (1959). *J. exp. Med.* **110**, 49–64.
Berman, L. (1963). *Blood* **21**, 246–249.
Berman, L. (1966). *Lab. Invest.* **15**, 1084–1099.
Bessis, M. D. (1961). *Lab. Invest.* **10**, 1040–1067.
Bloom, W. and Bloom, M. A. (1954). *In* "Radiation Biology" (A. Hollaender, ed.), Vol. 1, pp. 1191–1143. McGraw-Hill Book Co. Inc., New York.

Bosma, M. J., Makinodan, T. and Walburg, H. E. (1966). *Fedn Proc. Fedn Am. Socs exp. Biol.* **25**, 547.

Burns, W., Bauer, H., Paronetto, F., Abrams, G. D., Watkins, K. C. and Popper, H. (1964). *Fedn Proc. Fedn Am. Socs exp. Biol.* **23**, 547.

Campbell, D. H. and Garvey, J. S. (1963). *Adv. Immun.* **3**, 261–278.

Carter, D., Einheber, A. and Bauer, H. (1965). *Surg. Forum* **16**, 79–80.

Carter, D., Einheber, A., Bauer, H., Rosen, H. and Burns, W. F. (1966). *J. exp. Med.* **123**, 205–266.

Chatten, J. (1964). *Am. J. med. Sci.* **248**, 715–727.

Cohen, J. O., Newton, W. L., Cherry, W. B. and Updyke, H. (1963). *J. Immun.* **90**, 358–367.

Cohen, S., Vassalli, P., Benacerraf, B. and McCluskey, R. T. (1966). *Lab. Invest.* **15**, 1143–1155.

Cohn, Z. A. and Wiener, E. (1963). *J. exp. Med.* **118**, 1009–1020.

Dameshek, W. (1963). *Blood* **21**, 243–245.

De Bruyn, P. P. H. (1948). *Anat. Rec.* **101**, 373–405.

Doll, J. P. (1962). *Am. J. Physiol.* **23**, 291–295.

Dolowy, W. C. and Muldoon, R. L. (1964). *Proc. Soc. exp. Biol. Med.* **116**, 365–371.

Duhig, J. T. (1965). *Fedn Proc. Fedn Am. Socs exp. Biol.* **24**, 491.

Edwards, J. L. and Klein, R. E. (1961). *Am. J. Path.* **38**, 437–453.

Ekstedt, R. D. (1964). *J. exp. Med.* **120**, 795–804.

Fagraeus, A. (1960). *In* "Cellular Aspects of Immunity" (G. E. W. Wostenholme and M. O'Connor, eds), pp. 3–5. J. A. Churchill Ltd., London.

Fauve, R. M. (1964). *Revue fr. Étud. clin. biol.* **9**, 1, 100–105.

Fishman, M., Hammerstrom, R. A. and Bond, V. P. (1963). *Nature, Lond.* **198**, 549–551.

Fliedner, T. M., Kesse, M., Cronkite, E. P. and Robertson, J. S. (1964). *Ann. N.Y. Acad. Sci.* **113**, 578–594.

Ghossein, N. A., Azar, H. A. and Williams, J. (1963). *Am. J. Path.* **43**, 369–379.

Good, R. A., Dolmasso, A. P., Martinez, C., Archer, O. K., Pierce, S. C. and Papermaster, B. W. (1962). *J. exp. Med.* **116**, 773–795.

Gordon, H. A. (1960). *Am. J. dig. Dis.* **5**, 841–852.

Gowans, J. L. and Knight, E. S. (1964). *Proc. R. Soc. B* **159**, 257–282.

Gowans, J. L. and McGregor, D. D. (1965). *Prog. Allergy* **9**, 1–78.

Grabar, P. (1965). *Tex. Rep. Biol. Med.* **23** (suppl. 1), 278–284.

Gustafsson, B. E. (1959). *Int. Congr. Microbiol. VII Stockholm. Symposia*, 327–335.

Gustafsson, B. E. and Laurell, C. B. (1959). *J. exp. Med.* **110**, 675–684.

Heise, E. R. and Myrvik, Q. N. (1966). *Fedn Proc. Fedn Am. Socs exp. Biol.* **25**, 439.

Hobby, G. L., Lenert, T. F., Maier-Engallena, J., Wakely, C., Keblish, M., Monty, A. and Auerbach, O. (1966). *Am. Rev. resp. Dis.* **93**, 396–410.

Horowitz, R. E., Bauer, H., Paronetto, F., Abrams, G. D., Watkins, K. C. and Popper, H. (1964). *Am. J. Path.* **44**, 747–761.

Hudson, J. A. and Luckey, T. D. (1964). *Proc. Soc. exp. Biol. Med.* **116**, 628–631.

Hvros, A. G., Cali, A. and Azar, H. A. (1966). *Am. J. Path.* **48**, 627–639.

Ikari, N. W. (1964). *Nature, Lond.* **202**, 879–881.

Karetzky, M. and Rudolf, L. E. (1964). *Surgery Gynec. Obstet.* **119** 129–138.

Kim, Y. B., Bradley, G. and Watson, D. W. (1966). *Fedn Proc. Fedn Am. Socs exp. Biol.* **25**, 547.

Leblond, C. P. and Sainte-Marie, G. (1960). *In* "Haemopoiesis", pp. 152–183, Ciba Foundation Symposium.

Lerner, E. M. (1964). *Fedn Proc. Fedn Am. Socs exp. Biol.* **23**, 286.

Little, J. R., Brecher, G., Bradley, T. R. and Rose, S. (1962). *Blood* **19**, 236–242.

Luckey, T. D. (1963). "Germ-free Life and Gnotobiology". Academic Press, New York.

McLaughlin, M. M., Dacquisto, M. P., Jacobus, D. P., Horowitz, R. E. and Levenson, S. M. (1961). *Radiat. Res.* **14**, 484–486.

McMillan, D. B. and Engelbert, V. E. (1963). *Am. J. Path.* **42**, 315–335.

Martin, L. K., Einheber, A., Porro, R. F., Sadun, E. H. and Bauer, H. (1966). *Milit. Med.* **131**, 870–896.

Mellors, R. C. and Korngold, L. (1963). *J. exp. Med.* **118**, 387–396.

Meneghelli, V. (1961). *Acta anat.* **47**, 164–182.

Michael, J. G., Whitby, J. L. and Landy, M. (1962). *J. exp. Med.* **115**, 131–146.

Miller, J. F. A. P., Marshall, A. H. E. and White, R. G. (1962). *Adv. Immun.* **2**, 111–162.

Miller, J. F. A. P. (1963). *Br. med. J.* **2**, 459–464.

Miller, J. J. and Nossal, G. J. V. (1964). *J. exp. Med.* **120**, 1075–1086.

Miyakawa, M. N., Isomura, N., Shirasawa, H. and Yokoi, K. (1958). *Acta path. jap.* **8**, 79–97.

Moe, R. E. (1963). *Am. J. Anat.* **112**, 311–317.

Moe, R. E. (1964). *Am. J. Anat.* **114**, 341–370.

Murray, R. G. and Woods, P. A. (1965). *Anat. Rec.* **150**, 113–128.

Nossal, G. J. V. (1959). *Aust. J. exp. Biol. Med. Sci.* **37**, 499–504.

Nossal, G. J. V. (1962). *Intern. Rev. exp. Path.* **1**, 1–72.

Nossal, G. J. V. (1964). *Ann. N.Y. Acad. Sci.* **120**, 171–181.

Olson, G. B. and Wostmann, B. S. (1965). *Fedn Proc. Fedn Am. Socs exp Biol.* **24**, 581.

Olson, L. D. and Burnstein, J. (1965). *Fedn Proc. Fedn Am. Socs exp. Biol.* **24**, 371.

Oort, J. and Turk, J. L. (1965). *Br. J. exp. Path.* **46**, 147–154.

Osoba, D. (1965). *J. exp. Med.* **122**, 633–650.

Pernis, B., Cohen, M. W. and Thorbecke, G. J. (1963). *J. Immun.* **91**, 541–552.

Pleasants, J. R., Olson, G. B., Reddy, B. S. and Wostmann, B. S. (1964). *Fedn Proc. Fedn Am. Socs exp. Biol.* **23**, 408.

Pollard, M. (1964a). *Fedn Proc. Fedn Am. Socs exp. Biol.* **23**, 593.

Pollard, M. (1964b). *Science, N.Y.* **145**, 247–251.

Reed, N. D. and Jutila, J. W. (1965). *Science, N.Y.* **150**, 356–357.

Roberts, J. C., Jr. (1960). *In* "The Lymphocyte and Lymphocytic Tissue" (J. W. Rebuck, ed.), pp. 82–98. Paul Hoeber, Inc., New York.

Rovin, S., Costich, E. R., Fleming, J. E. and Gordon, H. A. (1965). *Arch. Path.* **79**, 641–643.

Rovin, S., Costich, E. R., Flemming, J. E. and Gordon, H. A. (1966). *J. oral Surg.* **24**, 239–246.

Salomon, J. C. (1965). *C. r. hebd. Séanc. Acad. Sci., Paris* **260**, 4862–4864.

Salvin, S. B., Peterson, R. D. A. and Good, R. A. (1965). *Fedn Proc. Fedn Am. Socs exp. Biol.* **24**, 160.

Schaffer, J. (1963). *Proc. Soc. exp. Biol. Med.* **112**, 561–564.

Schlesinger, M. and Mark, R. (1963). *Science, N.Y.* **143**, 965–966.

Schooley, J. C. (1961). *J. Immun.* **86**, 331–336.

Sell, S. (1965). *J. Immun.* **95**, 300–305.

Sherman, J. D., Adner, M. M. and Dameshek, W. (1963). *Blood* **22**, 252–271.

Sherman, J. D. (1965). *Fedn Proc. Fedn Am. Socs exp. Biol.* **24**, 491.

Skarnes, R. C. and Watson, D. W. (1957). *Bact. Rev.* **21**, 273–294.

Springer, G. F. (1959). *Z. Immunitätsforschung* **118**, 228–245.

Sprinz, H., Kundel, D. W., Dammin, G. J., Horowitz, R. E., Schneider, H. and Formal, S. B. (1961). *Am. J. Path.* **39**, 681–695.

Stollerman, G., Ekstedt, R. D. and Cohen, I. R. (1965). *J. Immun.* **95**, 131–140.

Suter, E. and Kirsanow, E. M. (1962). *Nature, Lond.* **195**, 397–398.

Swartzendruber, D. C. (1966). *Am. J. Path.* **48**, 613–619.

Tennant, R., Parker, J. C. and Ward, T. G. (1965). *J. natn. Cancer Inst.* **34**, 381–387.

Thorbecke, G. J. (1959). *Ann. N.Y. Acad. Sci.* **78**, 237–246.

Thorbecke, G. J., Benacerraf, B. (1959). *Ann. N.Y. Acad. Sci.* **78**, 247–254.

Thorbecke, G. J. and Benacerraf, B. (1962). *Prog. Allergy* **6**, 559–598.

Trowell, O. A. (1961). *Int. J. Radiat. Biol.* **4**, 163–173.

Vaughan, R. B. (1965). *Br. J. exp. Path.* **46**, 71–81.

Vazquez, J. J. (1961). *Lab. Invest.* **10**, 1110–1125.

Wagner, M. (1959). *Ann. N.Y. Acad. Sci.* **78**, 261–271.

Waksman, B. H., Arnason, B. G. and Jankovic, B. D. (1962). *J. exp. Med.* **116**, 187–206.

Ward, P. A., Johnson, A. G. and Abell, M. R. (1959). *J. exp. Med.* **109**, 463–474.

Weiss, L. (1964). *Bull. Johns Hopkins Hosp.* **115**, 99–174.

Weissmann, G. (1965). *New Engl. J. Med.* **273**, 1084–1143.

Welsh, R. A. (1962). *Am. J. Path.* **40**, 285–296.

Wensinck, F. (1961). *J. Path. Bact.* **81**, 395–400.

White, R. G. (1960). *In* "Mechanisms of Antibody Formation". Proceedings of a Symposium held in Prague, May 27–31, 1959. (M. Holub and L. Jarošková, eds), pp. 25–29. Academic Press, New York.

Wiener, J., Spiro, D. and Zunker, H. O. (1965). *Am. J. Path.* **47**, 723–764.

Wilson, R. (1963). *Radiat. Res.* **20**, 477–483.

Wilson, R., Sjodin, K. and Bealmear, M. (1964). *Proc. Soc. exp. Biol. Med.* **117**, 237–239.

Wilson, R., Bealmear, M. and Sobonya, R. (1965). *Proc. Soc. exp. Biol. Med.* **118**, 97–99.

Wohlwill, F. J. and Jetter, W. W. (1953). *Am. J. Path.* **29**, 721–729.

Wostmann, B. S. and Olson, G. B. (1964). *J. Immun.* **92**, 41–48.

Yoffey, J. M., Reinhardt, W. O. and Everett, N. B. (1961). *J. Anat.* **95**, 293–299.

Yoffey, J. M. (1964). *Ann. N.Y. Acad. Sci.* **113**, 867–886.

Chapter 11

Carcinogenesis in Axenic[1] Animals

JEAN-CLAUDE SALOMON

Institut de Recherches Scientifiques sur le Cancer,
C.N.R.S., Villejuif, (Val de Marne) France

The rapid extension of work on viral carcinogenesis during the last fifteen years has forced cancer workers to the idea that many experimental cancers are of infectious origin. Most tumour-responsible viruses exist in a latent state among wild mice as well as among laboratory mice. The idea of using techniques of axenic life in cancer research naturally followed.

Certain questions which arose at the beginning have already been at least partially answered.

(i) Do axenic animals develop spontaneous neoplasms?

(ii) Do the known carcinogenic agents, such as chemical substances, ionizing radiations and viruses, produce tumours in axenic animals? If so, are there differences between axenic and holoxenic animals in incidence, latency and histology of the tumours and their evolution?

(iii) Are the experimentally induced tumours in axenic animals transplantable and do they maintain their characteristics on serial passage?

In other words, is the germ-free animal a valuable tool in cancer research? The following review will show that present information is still fragmentary and will try to indicate some lines of research in which the use of axenic animals seems to be justified.

[1] Following the recommendations of Raibaud *et al.* (1966), the words *axenic* and *holoxenic* have been used in this chapter to describe germ-free and conventional animals, respectively.

I. Spontaneous Tumours in Axenic Animals

It is necessary to consider separately tumours occurring in aged rodents and tumours or leukaemias specific to certain strains of mice in which the high frequency is the principal distinctive characteristic of the strain.

Pollard and Teah (1963), observing axenic rats for a long time, report twenty-five spontaneous tumours in random bred Wistar rats of which nineteen were mammary tumours; of these, six were identified as adenofibromas, two as adenocarcinomas and eleven were not examined histologically. Three rats had thymic lymphosarcomas without metastasis and one rat had an interscapulary fibrosarcoma. Two more tumours in this series were less precisely described. The age of the animals varied from 13 to 38 months and, with the reservation that the number of animals was small, the distribution was identical whatever the number of generations of axenic animals from which the tumour-bearing rats were derived. No tumour was observed among Fischer and Sprague Dawley rats of their axenic colony. This may not be significant since they had been maintained in the germ-free state for a shorter period (2–3 years) than the Wistar rats (8 years).

Gordon et al. (1966) described naturally-occurring deaths of axenic mice derived from one holoxenic Swiss Webster pair. "In the course of post-mortem examination the impression was gained that the incidence of neoplasia, often affecting the lungs and the female reproductive organs, was at least as frequent among germ-free mice as it was among conventional controls". Mean life span, both for males and females, was very significantly longer in axenic than in holoxenic mice of the same strain. This immediately leads to the question of the effects of age on the incidence of spontaneous tumours. From the limited data it appears either that the real onset of cancer is later in axenic animals or that the progression of tumours is slower; both are essential points that deserve more clarity. The spontaneous character of these tumours is only the reflection of our ignorance about the mechanism of carcinogenesis. The accumulation of this type of data will help to determine a base level for the incidence of the so-called spontaneous tumours but it may only be a temporary level because the degree of environment control that can be attained with axenic methods is still very far from perfect. In particular the control of bacterial, fungal and viral floras (as yet not fully achieved) must be accompanied by the development of diets more certainly free from carcinogenic agents.

Pollard et al. (1965) described the occurrence of spontaneous leukaemia in axenic AKR mice, thus confirming previous studies of Gross (1961) and of Dmochowski et al. (1963) who found responsible viral particles of this leukaemia in foetuses. Furthermore Pollard (1967a) has observed in the Lobund colony of axenic AKR mice the occurrence of a complex syndrome; a progressive wasting in young adult mice is associated with thymic atrophy, enlargement of the lymph nodes and a disruption of the spleen morphology. No similar syndrome was detected in holoxenic AKR mice. By analogy with the description of Siegler and Rich (1963) of precocious stages of leukaemia, Pollard suggests an unmasking of a preleukaemic state which becomes clinically apparent in germ-free conditions.

Observations on axenic C3H mice derived from a strain with a high frequency of mammary tumours gave contradictory results with different authors. Reyniers and Sacksteder (1958) did not observe any mammary tumour in their series. Conversely, Kajima and Pollard (1965) noted the late appearance of mammary adenocarcinoma in axenic C3H mice with type B viral particles. Furthermore, Pilgrim and Labrecque (1967) found that the incidence of mammary tumours was virtually identical in their axenic C3H population and its holoxenic counterpart. Maybe the contradiction is only apparent since some authors dissociate the mammary tumours into two types according to the responsible viral agent, the mode of transmission, the type of tumour and the date of its appearance (Pitelka *et al.*, 1964).

II. Experimental Tumours in Axenic Animals

A. CHEMICALLY-INDUCED TUMOURS

In general, tumours appear in axenic animals given chemical carcinogens in the same way as they do in conventional animals. Transplantable fibrosarcomas have been induced by subcutaneous inoculation of 3-methylcholanthrene in adult mice of the Swiss Webster, Ha/ICR, CFW and C3H strains, in Fischer and Wistar rats (Pollard *et al.*, 1964) as well as in Leghorn and Bantam chickens (Reyniers and Sacksteder, 1959). Intra-gastric administration of a single dose (20 mg) of 7,12-dimethylbenzanthracene in sesame oil induced the appearance of mammary tumours in female Sprague Dawley rats (Pollard, 1963).

Susceptibility of the new-born to chemical carcinogens has been demonstrated in axenic animals. It was possible to induce the precocious appearance of lung adenomas in Swiss mice by a single neonatal subcutaneous inoculation of 3-methylcholanthrene (Pollard and Salomon, 1963). With C57/Bl new-born axenic mice, the repeated administration of urethane by subcutaneous route induced the formation of thymic lymphoma (Salomon, unpublished).

TABLE I

Effects of chemical carcinogens in axenic animals

Species	No. of survivors	Age	Sex	Carcinogen	Route	Dose (mg)	Type of tumour	Incidence	Reference
Mouse									
Swiss Webster	47	1 day	Both	3MC[b]	Subcutaneous	1 × 0·1	Lung adenoma	100%	Pollard and Salomon (1963)
Swiss Webster[a] ICR CFW C3H	134	Various	Both	3MC	Subcutaneous	1 × 0·5	Fibrosarcoma Epidermoid carcinoma	76% 2%	Pollard et al. (1964)
C57/Bl6	8	1 day and each week for 7 weeks	Both	Urethane	Subcutaneous	8 × 1·0/g	Thymic lymphoma	50%	Salomon (unpublished)
Rat									
Wistar Fischer[a]	41	Various	Both	3MC	Subcutaneous	1 × 1·0	Fibrosarcoma	90%	Pollard et al. (1964)
Sprague-Dawley	9	50–60 days	Both	DMBA[c]	Intragastric	1 × 20	Mammary adeno-carcinoma	100%	Pollard (1963)
Chicken									
White Leghorn[a]	30	17 days	?	3MC	Subcutaneous	1 × 5	Fibrosarcoma	87%	Reyniers and Sacksteder (1959)
Black Bantam	24	15 days	?	3MC	Subcutaneous	1 × 5	Fibrosarcoma	71%	

[a] Tumours arising in these animals were serially passaged successfully.
[b] 3MC = 3-methylcholanthrene.　　[c] DMBA = dimethylbenzanthracene.

From these observations, which are summarized in Table I, the incidence of tumours and time of their development seems in all ways similar to that usually described in holoxenic animals. There is no evidence for stating with any certainty that the existence of a complex flora influences tumorigenesis.

With orally active carcinogens the situation may be different. Cycasin, a glycoside derived from the nut *Cycas circinalis L.*, produces neoplasms of the kidney, intestines and liver and is highly toxic in large doses (Laqueur, 1964). However, it is much less toxic when given intra-peritoneally and there is little doubt that the toxicity and carcinogenicity are attributable to the aglycone released by the action of β-glucosidase in the gut. When cycasin is given by mouth to axenic rats, almost all is excreted unchanged in the faeces and neither toxic signs nor neoplasia are observed (Spatz *et al.*, 1966). Thus it must be assumed that bacterial action in the gut of the holoxenic rat brings about the conversion of the glycoside to cycasin aglycone.

B. CARCINOGENESIS BY IONIZING RADIATION

Some experiments on carcinogenesis by ionizing radiation have been done with axenic Swiss C3 Hand C57/Bl mice. Month-old mice received four times at weekly intervals a dose of 150 R by total exposure (Pollard and Matsuzawa, 1964). Four months after the last irradiation they developed lymphoid leukaemias with a predominant thymic localization. Electron microscope examination of the huge thymuses revealed the presence of viral particles of Bernhard C type.

Walburg *et al.* (1965), by irradiation of ICR axenic mice at a high single dose (600 to 900 R) or by irradiation of RF axenic mice with a single dose (300 R), obtained lymphoid leukaemias of which about half affected the thymus either exclusively or in a generalized lymphomatosis. The comparison of cumulative incidence of leukaemias at the tenth month showed a slightly higher incidence with axenic mice. As in holoxenic mice, the axenic mice that died from leukaemia had lesions of glomerulosclerosis.

Sprague-Dawley, Wistar and Fischer rats irradiated with a divided dose (Pollard and Kajima, 1966) did not develop leukaemia. However, six months after irradiation, two rats showed an increase in volume of the thymus, which was hard and fibrous, as well as an atrophy of lymph nodes. Between 9 and 12 months after irradiation, six out of forty-nine Wistar and six out of forty-three Sprague-Dawley rats developed mammary adenofibromas. One Sprague-Dawley and one Fischer rat out of twelve had mammary adenocarcinomas whereas no rat of the Sprague-Dawley or Fischer strains had ever had tumours in axenic life.

C. VIRAL CARCINOGENESIS

One point seems to be well established at present; by electron microscope examination, axenic mice have regularly been found harbouring viral particles of type C identical to particles responsible for murine leukaemias (de Harven, 1964). Particles of the same type were observed in the thymuses of axenic mice carrying spontaneous leukaemias in Ha/ICR and RF strains (Kajima and Pollard, 1965; Walburg et al., 1965). Although in this last case no formal evidence of the oncogenic role of these viruses has been demonstrated, one may reasonably suppose by analogy with leukaemic holoxenic mice that these particles are truly responsible for the leukaemias in axenic animals.

Different groups of workers have tested the effects of some oncogenic viruses on axenic animals. Mirand and Grace (1963) obtained typical Friend disease by intraperitoneal inoculation of an acellular filtrate prepared from the spleen of a conventional mouse suffering from Friend leukaemia. At the 14th and 21st days axenic ICR mice had a comparatively much more severe disease, as judged by a higher splenic weight, a bigger erythropoietic activity and a greater leucocytosis. These findings demonstrate that the neoplasia produced by the Friend virus in axenic ICR mice was quicker in onset or more pronounced, or both, than in holoxenic mice.

Pollard (1964), by inoculation of strain A (Gross) leukaemic virus to axenic new-born C3H mice, caused a typical leukaemogenesis of which the first histological signs appeared after 3 weeks. Clinically detectable leukaemias were observed at the ninth week. As in holoxenic animals, there was an increase in volume of the thymus and lymphatic glands. The spleen, kidneys and liver were pale and infiltrated with lymphoid cells.

Pollard and Kajima (1966), by intraperitoneal inoculation of Sprague-Dawley axenic rats with Gross virus strain A adapted to the rat, induced leukaemias in almost all animals at about 90 days after birth. These lymphoid leukaemias, which evolved fatally very rapidly after the first symptoms, again involved the thymus and less consistently the other lymphopoietic organs infiltrated with lymphoblastic cells. Type C particles, $80{-}100 \, \mu$ in diameter, could be seen in the tumoral cells of the thymus.

Reyniers and Sacksteder (1958, 1959) using Rous sarcoma virus of the Bryan strain were able to induce the formation of sarcoma in Leghorn chickens.

III. Immune Response in Axenic Animals During Carcinogenesis

Pollard has examined histologically the lymphoid organs in axenic AKR mice (Pollard et al., 1965; Pollard, 1964) during spontaneous or experimental carcinogenesis (Pollard, 1967b). In young AKR mice, before the appearance of evident signs of leukaemia, there is an enlargement of lymph nodes and the germinal centres are small and spread. Medullary zones contain clumps of large reticular cells and relatively few lymph cells; this is different from the appearance in axenic mice of other strains. At a later stage, at the second or third month of age during the preleukaemic phase, these medullary zones become rich in cells like plasmocytes. Within the same period the spleen undergoes important structural disorders: white and red pulps are badly outlined and the red pulp is infiltrated by large lymphoid cells, plasma cells and reticular cells which seem to have migrated from the perifollicular envelopes.

In primitive tumours of axenic rats, there is an accumulation of plasma cells in lymphoid organs, an increase in the number of lymphocytes in the red pulp but no germinal centres. The immunoglobulin remains at a low level.

On the whole, from these findings it appears that the immunological behaviour of axenic animals does not differ essentially from that of holoxenic animals, but that the weakness of previous antigenic stimuli allows a better definition of the factors concerned with the carcinogenesis itself.

IV. Prospects

It is clear enough that there is no need in the future to repeat all fundamental cancer experiments on axenic animals. Work already published shows that there is little or no essential difference between holoxenic and axenic animals in their response to tumorogenic agents. The choice of further experiments will be based on several considerations. Although very soon we should know with more confidence if axenic rats are really free of latent virus, it will certainly be more difficult to get mice free from all contamination with type C virus particles. The phenomenon of transmission of virus from parent to offspring is still not clear. It is not known (a) if the virus(es) is (are) transmitted as an infectious complete particle through the placental barrier, (b) if there is an infection of the germ cells or (c) if the viral genome is integrated in the germ cells. It would be unreasonable to expect axenic techniques to clear a strain of mouse from one or several satellite viruses of this type. Only those viruses which infect the baby mouse after birth can be eradicated.

In some cases the use of axenic animals is indispensable as, for example, (a) in studies of carcinogenic agents metabolized by the digestive flora or by one of its products, as has been shown for cycasin or (b) a study of the effects on carcinogenesis of thymectomy in the newborn, where it seems that it is not the axenic state *stricto sensu* that protects animals from the wasting syndrome but the absence of certain still undefined infectious agents (see also Chapter 10).

In other circumstances the decision to use axenic, in preference to "clean", animals will be advisable, as in the following instances: (a) When it is necessary to use cytostatic agents or strong immunodepressors such as corticoids, the fear of lethal infection completely justifies the choice of axenic animals. (b) Every time that the study of an immune process, or of the reaction of the reticuloendothelial system to carcinogenesis, is likely to be masked by a background of responses to non-specific stimuli in an unstable environment of complex infections. (c) In the collection of more basic data on the so-called spontaneous carcinogenesis of the rat and mouse, and also of the chicken on which there has been no systematic study, the use of axenic animals will help to ensure that no inapparent infection is responsible for the phenomenon. (d) Similarly, the mechanism of spontaneous malignant transformation in *in vitro* cell lines can best be studied in cells derived from axenic animal tissues.

For the future, the introduction of isolation methods and the possibility of obtaining breeding stocks deprived both of viruses and the usual pathogenic bacteria, ideas that are gradually finding their way into cancer research, are directly derived from germ-free methodology and studies with gnotobiotic animals.

REFERENCES

de Harven, Et. (1964). *J. exp. Med.* **120**, 857.

Dmochowski, L., Grey, C. E., Padgett, F. and Sykes, J. A. (1963). *In* "Viruses, Nucleic Acids and Cancer". *Proc. 17th Ann. Symp. on Fundamental Cancer Research*, University of Texas, M. D. Anderson Hospital and Tumor Inst., p. 85. Williams and Williams Co., Baltimore, Maryland, U.S.A.

Gordon, H. A., Bruckner-Kardoss, E. and Wostmann, B. S. (1966). *J. Geront.* **21**, 380–387.

Gross, L. (1961). "Oncogenic Viruses." Pergamon Press, New York.

Kajima, M. and Pollard, M. (1965). *J. Bact.* **90**, 1448.

Laqueur, G. L. (1964). *Fedn Proc. Fedn Am. Socs exp. Biol.* **23**, 1386–1388.

Mirand, E. A. and Grace, J. T., Jr. (1963). *Nature, Lond.* **200**, 92–93.

Pilgrim, H. I. and Labrecque, A. D. (1967). *Cancer Res.* **27**, 584–586.

Pitelka, D. R., Bern, H. A., Nandi, S. and De Ome, K. B. (1964). *J. natn. Cancer Inst.* **33**, 867.

Pollard, M. (1963). *Nature, Lond.* **200**, 1289.

Pollard, M. (1964). *Science, N.Y.* **145**, 247–251.

Pollard, M. (1965). *Prog. med. Virol.* **7**, 362–376. Karger, Basel-New York.

Pollard, M. (1967a). *Nature, Lond.* **213**, 142–144.

Pollard, M. (1967b). *Fedn Proc. Fedn Am. Socs exp. Biol.* **26**, 625.

Pollard, M. and Kajima, M. (1966). *Proc. Soc. exp. Biol. Med.* **121**, 585–589.

Pollard, M. and Matsuzawa, T. (1964). *Proc. Soc. exp. Biol. Med.* **116**, 967–971.
Pollard, M. and Salomon, J. C. (1963). *Proc. Soc. exp. Biol. Med.* **112**, 256–259.
Pollard, M. and Teah, B. A. (1963). *J. natn. Cancer Inst.* **31**, 457–465.
Pollard, M., Kajima, M. and Teah, B. A. (1965). *Proc. Soc. exp. Biol. Med.* **120**, 72–75.
Pollard, M., Matsuzawa, T. and Salomon, J. C. (1964). *J. natn. Cancer Inst.* **33**, 93–99.
Raibaud, P., Dickinson, A. B., Sacquet, E., Charlier, H. and Mocquot, G. (1966). *Ann. Inst. Pasteur* **111**, 193–210.
Reyniers, J. A. and Sacksteder, M. R. (1958). *Proc. Anim. Care Panel* **8**, 41–54.
Reyniers, J. A. and Sacksteder, M. R. (1959). *Ann. N.Y. Acad. Sci.* **78**, 328–353.
Siegler, R. and Rich, M. A. (1963). *Cancer Res.* **23**, 1669.
Spatz, M., McDaniel, E. G. and Laqueur, G. L. (1966). *Proc. Soc. exp. Biol. Med.* **121**, 417–422.
Walburg, H. E., Jr., Upton, A. C., Tyndall, R. L., Harris, W. W. and Cosgrove, G. E. (1965). *Proc. Soc. exp. Biol. Med.* **118**, 11–14.

Chapter 12

Radiation Biology

D. W. VAN BEKKUM

Radiobiological Institute TNO,
Rijswijk (ZH), The Netherlands

I. Introduction

The most important contributions of gnotobiotics to radiation biology
have thus far been made in the field of mammalian radiation disease.
Among the tissues most susceptible to radiation damage are the lymphatic
organs, so that the immunological defence of irradiated organisms will be
impaired. Lethally irradiated animals are therefore highly susceptible to
infections with microorganisms that do not normally cause disease. Such
infections interfere with many studies involving exposure of the animals
to high doses of whole body irradiation by causing confusion as to the
specificity of the various organ lesions encountered and by adversely
influencing the survival time of the irradiated animals. The elimination
of these disturbing factors may sometimes, but not always, necessitate the
removal of the entire microflora before the irradiation. This review will
concentrate on the use that has thus far been made of gnotobiotes to

elucidate the role of microorganisms in determining the responses of experimental animals to whole body irradiation.

Following exposure of mammals to high doses of whole body irradiation with ionizing radiation, highly characteristic clinical syndromes develop, depending on the dose received. For those readers who are not familiar with dose-survival relationships, a few introductory remarks will be made. For a more detailed treatise the reader is referred to the recent monographs by Bond *et al.* (1965) and by van Bekkum and de Vries (1967).

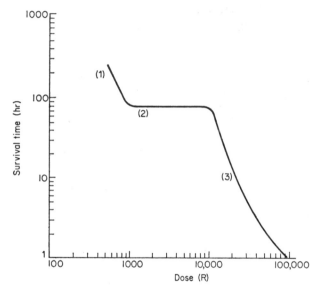

FIG. 1 Relation between dose of whole body irradiation and survival time in mice. (1) Bone marrow syndrome, (2) gastro-intestinal syndrome, (3) cerebral syndrome.

The symptomatology, as well as the survival time, following a single whole body irradiation is determined by the radiation dose. In mammals three different radiation syndromes can be distinguished (Fig. 1).

Following doses in excess of about 12,000 R[1] signs characteristic of central nervous system damage — such as convulsions, paralysis and stupor — occur within hours and increase in severity until death ensues. The survival time varies between a few hours and 1 or 2 days and is inversely dependent on the magnitude of the dose. This form of radiation disease is known as the *cerebral syndrome*; it can be prevented by shielding the head and conversely it can be provoked by delivering such excessively

[1] Röntgen (R) will be used as the unit of dose of ionizing radiation throughout this review because it has been employed in most of the literature reviewed.

high doses to the head only, when the body of the animal is being shielded. Histopathological changes which have been found to accompany the cerebral syndrome consist of perivascular and parenchymal granulocytic infiltrates in the meninges, the choroid plexuses and the brain, occuring as early as 2 hr following irradiation. Vasculitis and œdema involving all tissues of the central nervous system (CNS) develop within 24 hr, sometimes as early as 3 hr after irradiation. In addition, pyknosis and karyolysis of the granular neuronal cells of the cerebrum has been found in several species.

Until recently there was no evidence that bacteria play a role in the cerebral syndrome but the studies of McLaughlin *et al.* (1964) have revealed striking differences in the reactions of germ-free and conventional mice to exposures in excess of 15,000 R of whole body irradiation. These observations will be discussed more extensively in a later section.

In the dose region between roughly 1200 R and 12,000 R mortality is due to severe damage to the intestinal tract and its consequences. Within a few days the animals develop diarrhoea, which is more pronounced in some species than in others. Mice and rats die between the 3rd and 5th day following irradiation and death is associated with a disturbance of the water and mineral balance resulting from impairment of the reabsorption of water and sodium from the intestinal lumen (Curran *et al.*, 1960). The diarrhoea is viscous in rodents but excessive and watery in dogs, monkeys and humans and involves severe losses of protein.

Sullivan *et al.* (1965) showed that diarrhoea does not occur in heavily irradiated rats when the bile is directed away from the intestinal lumen by cannulation of the bile duct. They postulate that bile removes the mucus from the intestinal epithelium of the irradiated rat and suggest that mucus may be an essential part of the permeability barrier which normally prevents loss of water and salts into the intestinal lumen. The intestinal syndrome can be prevented to some extent by shielding of the intestines; the protection of even a small portion of the ileum has been found to be effective (Swift and Taketa, 1956; Smith, 1960), indicating that a relatively small piece of functioning intestine may be sufficient to counteract the deficient resorption in the other parts of the gut. Proof that the excessive fluid loss is the cause of death was provided by experiments in dogs, in which a continuous replacement of fluid and minerals was effective in preventing the fatal outcome of the intestinal syndrome (Conard *et al.*, 1956).

The histological basis of the radiation damage to the intestinal tract has been provided by Quastler (1956) and Sherman and Quastler (1960). They found that the epithelial cells of the intestinal mucosa of mice are continuously moving from the crypts, where cell division occurs, to the top of the villi, where the cells are eventually shed (Quastler and Sherman, 1959). The renewal of the intestininal epithelial lining takes about 4 days; the movement of a cell along the villus takes 2·3 days in the mouse (Matsuzawa and Wilson, 1965). The high radiation doses which induce the intestinal syndrome cause a complete block of mitoses of the crypt cells so that the renewal of epithelial cells is abruptly discontinued. Those cells present in the crypts continue their movement to the top of the villi, consequently the number of epithelial cells covering the villi decreases rapidly and after 3·5 days only a few are left, leaving the villi denuded. Shrinkage of the villi occurs concurrently but is insufficient to compensate for the loss of epithelial cells. It is not difficult to envisage that this denudation of the villi is accompanied by a marked disturbance of the normal reabsorptive capacity of the intestinal tract.

Exposure to doses below roughly 1200 R produces the bone marrow syndrome which is dominated by deficient functions on the part of the haemopoietic system. The severity of the symptoms as well as the interval between irradiation and death are again determined by the dose of radiation. The essential cause of this form of radiation disease is the inhibition of the proliferation of haemopoietic and lymphopoietic cells. Accordingly, provided the dose of radiation was high enough to produce a nearly complete block of mitotic activity, the peripheral blood cells will disappear at a rate corresponding to their life span, to which the period of maturation from non-dividing precursors in the bone marrow has to be added.[1] Marked differences in the dose response and the time response of the peripheral blood picture have been observed between the various mammalian species but appear to be proportional to the species-specific cell renewal characteristics (Bond et al., 1965).

In the case of lymphoid cells there occurs in addition a direct killing effect on the "mature" small lymphocytes, resulting in so-called inter-phase or intermitotic death which could contribute the severe lympho-paenia seen within 24 hr following irradiation. Granulocytes are not themselves killed by radiation at this dose level but as their production is blocked, minimal values are reached after 3–4 days in mice following lethal doses of irradiation. In most species a marked granulocytosis occurs within a few hours after irradiation, presumably as a result of mobilization from the extravascular pools. Complete disappearance of thrombocytes from the peripheral blood usually occurs at 6–8 days after irradiation in the mouse, depending on the dose of irradiation and the occurrence of infections.

[1] This account of events is of necessity simplified. The picture is complicated by various factors, such as formation of giant cells, the functional capacities of which are problematic.

The life span of erythrocytes is much longer, 60 days in mice and rats and 120 days in man, so that early anaemia is primarily due to multiple haemorrhages.

The clinical picture of the full blown acute bone marrow syndrome therefore is a consequence of the manifestation of infections and haemorrhages followed later by anaemia if survival time is sufficiently long. Which of the symptoms predominates is determined by the animal species, the composition of its microflora and also to some degree by the dose of irradiation. In the mouse, bacteraemia is nearly always found at the time of death, while a significant proportion of rats and monkeys die from haemorrhage in the absence of bacteraemia. If so-called conditioned pathogenic microorganisms happen to be present in the microflora, infection occurs usually at an earlier time resulting in precocious death from septicaemia. For example, mice carrying *Pseudomonas aeruginosa* die between the 4th and 7th day following exposure to a LD_{100} dose of radiation, while in the absence of this microorganism the mean survival time is 10 days (Wensinck *et al.*, 1957).

A combination of factors may also lead to early death. For instance when submucous haemorrhage in the intestinal tract causes local necrosis and ulceration, such ulcers may promote penetration into the blood stream by intestinal microorganisms resulting in early septicaemia. The outstanding feature of the bone marrow syndrome is a markedly decreased resistance to infections of any kind. This may result in a variety of manifestations, according to the microorganisms present and the ports of entry provided to them. Evidently, germ-free and gnotobiotic animals offer unique possibilities of unravelling these various factors. It should be kept in mind that the decrease in resistance develops similarly after doses exceeding the range which causes the bone marrow syndrome, so that involvement of infections in the intestinal syndrome and the cerebral syndrome can by no means be excluded.

II. Radio-sensitivity of Germ-free Animals in Terms of Mortality and Survival Time

As early as 1956 Reyniers *et al.* (1956) reported that germ-free rats survived approximately twice as long as conventional ones following exposure to single doses of whole body irradiation between 300 and 1000 R. This abstract did not contain any specific information concerning the mode of death, except for the statement that " . . . the germ-free rat in general showed the same kind of symptoms after irradiation as the conventional rat, however with certain quantitative variation". Similar studies were performed with mice by Wilson and Piacsek (1962) and by

Wilson (1963) who exposed the animals to doses of X-radiation varying from 500 R to 3000 R. They used the non-inbred Swiss-Webster strain, which is maintained at Lobund Institute. These observations were later repeated and extended with the same mouse strain to cover the dose range of 500–35,000 R by Matsuzawa (1965), also at Lobund. In 1964 McLaughlin *et al.* (1964) from the Walter Reed Army Institute of Research described their results after radiation doses between 950 R and 40,000 R, again using the Swiss-Webster strain of mice and recently

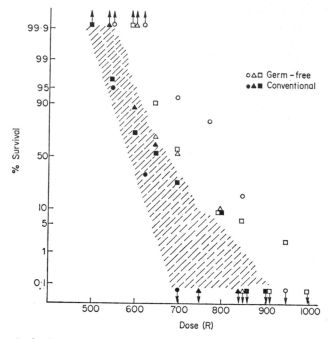

FIG. 2 Survival of conventional and germ-free mice following various doses of whole body irradiation. Combined data from McLaughlin *et al.* (1964), ○●; Wilson (1963), △▲ and Matsuzawa (1965), □■.

Walburg *et al.* (1966) at Oak Ridge reported on the responses of two non-inbred and one inbred strain of mice following exposure to doses between 500 and 1000 R.

The main results of these four studies show a very definite pattern in the differences between germ-free and conventional animals. In the dose range causing the bone marrow syndrome, the mortality curves of the germ-free mice show a shift to the right by about 100 R on the average. Figure 2 depicts the combined data of Wilson (1963), McLaughlin *et al.*

(1964) and Matzusawa (1965), who all worked with the Swiss-Webster strain. There is some difference in radiosensitivity between the mice from the two laboratories both in the conventional and in the germ-free animals. In each case, however, the germ-free mice were less sensitive than the conventional animals.

In Wilson's experiments the difference between the LD_{50} for the two groups was 45 R, in the experiments of McLaughlin this difference was 130 R, while Matzusawa found a difference of roughly 75 R. None of the individual survival curves has been determined at a sufficiently large number of points to provide very accurate information but the combined data leave no doubt that the germ-free mice are slightly more resistant to "haemopoietic" lethal damage.

TABLE I

LD_{50} values for germ-free and conventional mice from 3 different strains. Data from Walburg et al. (1966).

| Strain | Sex | LD_{50}: mean (95% confidence limits) in R | | |
		Germ-free	Conventional	Difference
ICR	♀	788 (769–806)	735 (710–761)	53
	♂	—	718 (686–747)	—
CF No. 1	♀	719 (705–733)	695 (634–756)	24
	♂	710 (697–722)	665 (583–746)	45
RFM/Un	♀	771 (758–785)	631 (616–647)	140
	♂	735 (719–751)	582 (562–598)	153

The experiments of Walburg et al. (1966) have yielded more accurate dose survival curves of germ-free and conventional mice for three different strains. Their LD_{50} values are listed in Table I, showing differences of 24 R to 150 R between the germ-free mice and their conventional controls. The mean survival time of the germ free mice following lethal haemopoietic injury (doses up to about 1000 R) was found to be greater than for conventional mice. In Fig. 3 the survival time is depicted in relation to the mortality level. There is no consistency in the pattern of these relationships, except that the survival time tends to decrease in germ-free as well as in conventional animals with increasing mortality levels. Survival times for germ-free mice vary between 24 and 12 days, for conventional animals between 16 and 6 days. The causes of these rather large variations between strains and laboratories are not known but may be supposed to be related to differences in food and its sterilization, radiation set-up and genetically determined differences in radio-sensitivity between the mouse strains. As regards the conventional mice, differences in the composition of the microflora may have been involved as well. The results thus far discussed fall in the first part of the dose survival curve, which represents the dose region in which the bone marrow syndrome is the cause of death. Very little information has been supplied in these

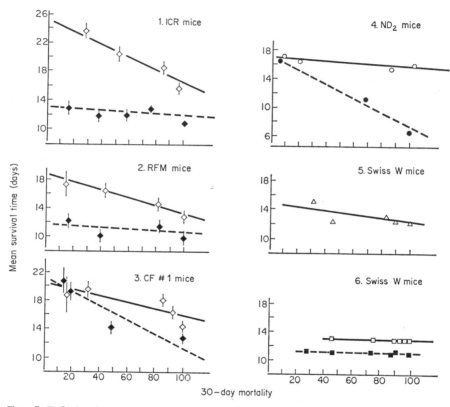

Fig. 3 Relation between survival time and per cent mortality in groups of whole body irradiated conventional and germ-free mice. Data from Walburg *et al.* (1966), graphs 1, 2, 3; McLaughlin *et al.* (1964), graph 4; Wilson (1963), graph 5; Matsuzawa (1965), graph 6. All open symbols indicate germ-free animals and all solid symbols conventional animals.

reports on the signs and the pathology of the animals so that the exact reasons for the prolonged survival of the germ-free animals and for their somewhat higher radio-resistance have not been established.

If it is accepted that conventional mice die in this dose region from septicaemia and haemorrhage, the results obtained with germ-free animals would indicate that the primary cause of death in the conventional animals is septicaemia and that prevention of this complication causes a slight protection at the lower lethal doses as well as an increased survival time by a few days in the lethally irradiated animals. This conclusion is supported by the finding of McLaughlin *et al.* (1964) that mice which were monoassociated with *E. coli* showed an identical dose mortality response as the conventional mice. On the other hand, the limited data on survival time of *E. coli* monoassociated animals suggest a slightly increased survival time (1–2 days), indicating that the type of microorganism present may be decisive for the severity and time of occurrence of the septicaemia. In other words, it seems possible that the presence of microorganisms in itself might not always lead to a fatal infection before death from haemorrhage occurs. It is noteworthy in this respect that Matsuzawa (1965) could not recognize any essential difference in the clinical and post-mortem appearances in this region (the first plateau) of the dose survival time curve. He stated specifically that the most pronounced pathological alteration was severe haemorrhage in subcutaneous, subserosal and submucosal tissues in both the germ-free and the conventional animals. Haematocrit values dropped to below 10% immediately before death in germ-free as well as in conventional mice, as a result of extensive haemorrhages. Clearly, studies with monoassociated animals will have to be greatly extended with other types of contaminants before the role of specific microorganisms in the post-irradiation mortality pattern can be properly evaluated. The possibility that haematological factors are responsible for the increased survival time of germ-free mice in the haemopoietic death zone, will be discussed in a subsequent paragraph.

The *second part* of the dose mortality curve has the form of a plateau representing the dose range in which gastrointestinal death occurs, begins at 1500 R in conventional and about 2000 R in germ-free mice (Fig. 4). In the transitional zone between the first and the second plateau the germ-free mice show extensive haemorrhages as well as gastrointestinal damage, while a third pathological entity, namely "enlarged liver and spleen with indications of septicaemia", has been observed in conventional animals by Matsuzawa (1965).

Much more interesting is the well established difference in survival time between 2000 R and 15–20,000 R; germ-free mice live twice as long (7–8 days) as conventional animals (3–4 days). Diarrhoea begins 2–3 days after irradiation in conventional mice and 5–6 days after irradiation in germ-free mice. In both types of animal there is extensive destruction of intestinal mucosa at the time of death.

There is no complete agreement with regard to survival times of the mice in this area between the various authors. The values given above refer to those of Matsuzawa (1965), whose data contribute most to the combined curve of Fig. 4. McLaughlin *et al.* (1964) report a difference of only 2 days between germ-free (7 days) and conventional mice (4–5 days). The latter survival time was also recorded for their *E. coli* monoassociated mice.

In Wilson's study a plateau of 3 days survival time was reached after 1700 R for conventional mice, while at the highest dose of 3000 R employed by him, the germ-free animals died after an average of 6·8

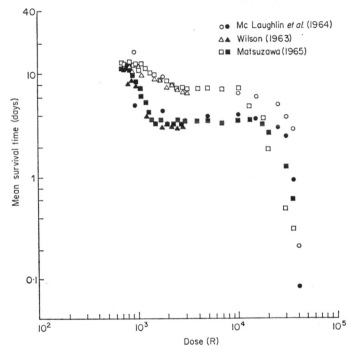

FIG. 4 Relation between survival time and radiation dose of germ-free and conventional mice exposed to whole body irradiation. Data from McLaughlin *et al.* (1964), ○●; Wilson (1963), △▲ and Matsuzawa (1965), □■. All open symbols indicate germ-free animals and all solid symbols conventional animals.

days but it is not certain that the survival time had reached a plateau at that dose. The investigations of Walburg *et al.* (1966) do not extend to doses above 1000 R so that the gastrointestinal death region was not approached in their studies.

Whatever the causes of these differences in the duration of the gastro-intestinal syndrome noted by the three groups of investigators, the prolonged survival of germ-free mice was generally confirmed. The most obvious explanation for this prolongation would seem to be the absence of microorganisms which can penetrate the denuded intestinal mucosa and overwhelm the animal. This explanation is however far too simple, since another major feature of germ-free mice is the smaller height of their intestinal villi and the fact that the rate of epithelial migration along the villi is much slower in the germ-free mouse. This implies that, following mitotic arrest in the intestinal crypts, it takes several days longer for the villi to become totally denuded. These cytological aspects of radiation damage to the gut will be fully discussed in another part of this chapter.

The information on the responses of germ-free mice to X-ray doses exceeding 20,000 R is limited and conflicting. Matsuzawa's data show a crossing of the survival time-dose graphs for conventional and germ-free mice at this point, the germ-free animals showing shorter survival times than the conventional ones. In this dose region death is presumed to occur as a result of damage to the central nervous system. Both types of animal showed an abnormal position of the body and continuous shivering during or after irradiation according to Matsuzawa but no attempt to explain the shorter survival time of the germ-free mice has been made.

The only other observations in this dose region are those of McLaughlin et al. (1964). They report survival times of 57 hr after 30,000 R, 22 hr after 35,000 R and 2 hr after 40,000 R for the conventional mice and 90 hr, 69 hr and 5 hr, respectively, for the germ-free animals. E. coli mono-associated mice were only studied at 40,000 R following which a survival time of 3 hr was noted. The distribution of the deaths following these extremely large doses of irradiation is shown in Fig. 5, which suggests a real difference in the responses of germ-free as compared to conventional mice.

McLaughlin et al. (1964) noted that the conventional animals exposed to 40,000 R, which took about 2 hr, suffered from convulsions and that their exposure cage was covered with bloody diarrhoea. The E. coli monoassociated mice showed convulsions and diarrhoea but did not excrete any blood. The germ-free animals had tremors but convulsions were only seen as a terminal event. There was no evidence of blood or diarrhoea in the exposure cages at any time. At radiation doses lower than 30,000 R, e.g. 15,000 and 25,000 R, the conventional mice showed evidence of central nervous system involvement, whereas these signs were absent in germ-free mice. These important observations have unfortunately not been substantiated by histopathological data so that one can only speculate as to the underlying mechanism of the strikingly different responses of germ-free animals. The authors pointed out that the higher threshold of germ-free animals for central nervous system damage might be due to a protective feature associated with the germ-free state or alternatively to a more rapid development of the disturbances of water and mineral metabolism in the conventional animals that might contribute to the manifestation of the central nervous system damage. A third possibility seems that the entrance of bacterial toxins into the

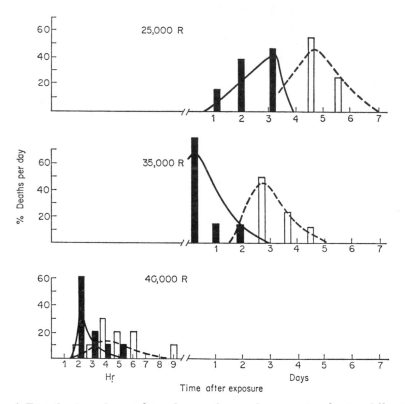

FIG. 5 Distribution of mortality of germ-free and conventional mice following whole body irradiation with doses of 25,000–40,000 R. Results from McLaughlin et al. (1964). □ = Germ-free, ■ = conventional animals.

circulation of the conventional animals results in an excitation of the central nervous system, which could have been sensitized by the radiation-induced damage.

The reverse ratio between survival times of conventional mice and germ-free mice after 35,000 R reported by Matsuzawa (1965) may be related to the lower dose rate (50–100 R/min) employed in his experiments as compared to 300–330 R/min in those of McLaughlin et al. (1964).

Evidently, the effects of the germ-free state on the responses of animals to doses causing irreversible damage to the central nervous system have to be studied in more detail. The results so far available merely indicate that the presence of a microflora may have a significant influence on these responses and it seems certain that such studies will contribute to our knowledge of the pathogenesis of the radiation-induced cerebral death.

The only other animal species reported as having been irradiated in the germ-free state is the chicken (McLaughlin *et al.*, 1958). The dose range between 350 R and 1850 R was studied using 10-day-old chickens. Following doses under 800 R the percentage survival of the germ-free birds was markedly higher than that of the conventional chickens. Above 800 R the survival time of the germ-free chickens was longer than that of the conventional birds. These results are reported in an abstract only, which does not contain more detailed information.

III. Haemopoiesis and the Bone Marrow Syndrome in Germ-free Mice

Wilson *et al.* (1964) suggested that normal germ-free mice have about 10% lower counts of peripheral white blood cells and reticulocytes, as well as a lower haematocrit as compared to conventional animals. Following whole body irradiation the germ-free mice showed a less severe and less rapid decrease of neutrophils at doses of 300–850 R. The anaemia which developed after 710 R was of a lesser degree and after 850 R its development was slower. Similarly, platelet counts were less reduced after 710 R in the germ-free mice but in the other radiation dose groups no significant differences were observed. The authors postulated that the bacterial flora is responsible for a greater "consumption" of granulocytes in conventional animals. As a consequence of their increased life span, the decrease of the leucocytes would be slower following a block of the mitotic activity of the myelopoietic cells. A similar dependence of thrombocyte survival on the presence of microflora seems less obvious but by no means impossible. Some relation between the thrombocytes and the presence of microorganisms is suggested by the well established clinical experience that thrombocyte survival is sharply decreased in patients with systemic infections. The slower development of anaemia in germ-free mice after irradiation might have been related to the pattern of development of thrombocytopenia but such an explanation seems to apply to the 710 R group only.

The rates of recovery of the various blood elements following sublethal irradiations were also recorded by Wilson *et al.* (1964) who found no significant differences.

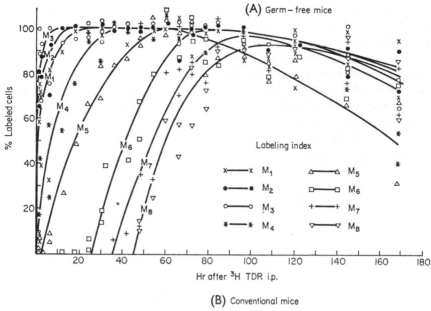

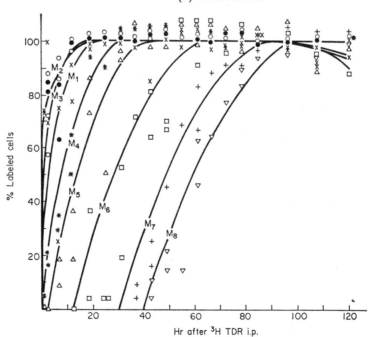

FIG. 6A and B Pattern of incorporation of tritiated thymidine (^{3}H TDR) in bone marrow cells of germ-free (A) and conventional (B) mice. M_1 = myeloblasts, M_2 = promyelocytes, M_3 = large myelocytes, M_4 = small myelocytes, M_5 = metamyelocytes, M_6 = juvenile neutrophils, M_7 = band-form neutrophils, M_8 = segmented neutrophils. From Fliedner *et al.* (1967).

Fliedner *et al.* (1967) have performed extensive studies on the proliferative kinetics of bone marrow cell renewal system in germ-free mice using conventional mice of the same NDZ-strain of the Radiobiological Institute in Rijswijk.

Tritiated thymidine (^{3}H-TDR) was injected intraperitoneally and animals were sacrificed at various time intervals for evaluation of the labelling pattern of bone marrow cells as a function of time. It was found (Fig. 6A and B) that the labelling index of granulocytic precursors initially after ^{3}H-TDR administration was similar in germ-free mice and their conventional controls. Further, there was no difference in the changes of the labelling indices in the various granulocytic precursor compartments after ^{3}H-TDR injection. It is of general interest that in both animal groups, the labelling index rose to practically 100% before 30 hr in cells termed M_1, M_2, M_3, M_4 (myeloblasts, promyelocytes, large and small myelocytes). No evidence was found for a difference in the proliferative rate of myelopoiesis in germ-free *vs* conventional mice. However, if the successive appearance of labelling was followed in the maturing-only compartments (i.e. metamyelocytes M_5, juveniles M_6, band-forms M_7 and segmented neutrophils M_8), it was found that the increase in the labelling indices was somewhat slower in germ-free mice. The time difference between labelling waves was in the order of 12 hr. This was also apparent when the first appearance and subsequent rise of labelled mature granulocytes was studied in blood smears. It was noted that the steepest rise of the labelling index of neutrophilic granulocytes was 10–12 hr later in germ-free as compared to conventional mice (Fliedner *et al.*, 1966). In conclusion, there does not seem to be a marked difference in cellular renewal in bone marrow cells — at least for myelopoiesis — except that neutrophilic granulocytes may mature somewhat slower in germ-free than in conventional animals. Eosinophilic granulocytes mature about 20 hr earlier than neutrophilic granulocytes, but there was no difference between germ-free and conventional animals (Fliedner *et al.*, 1966). Lymphocytes attain a maximal labelling index of about 10% in conventional and about 25% in germ-free mice after ^{3}H-TDR injection (Fliedner *et al.*, 1966). The authors conclude that this may reflect the size difference of the lymphocyte pools in germ-free and conventional animals.

When germ-free animals were given 700 R whole body irradiation (Teitge, 1967), no difference was found in the pattern of cellular reduction of myelopoiesis and erythropoiesis between germ-free and conventional mice. It was shown (Fig. 7) that the total number of myelocytic cells remains constant for about 1 day, due to the continued maturation of cells. Thereafter, a decrease occurs reaching minimal levels after about 4 days.

A small "abortive" rise was noted at day 7. There was no general difference in the slopes between germ-free and conventional mice. Although the cell numbers appeared to remain up between days 2 and 3 in germ-free mice, a sharp decrease was noted between days 2 and 3 in conventional mice. The numbers of animals used are too small to make a meaningful statistical analysis of these differences. However, they are reasonable on the basis of the slower maturation of granulocytes of germ-free mice, since in these animals, granulocytes would continue to be delivered into the blood for about 12 hr longer than in conventional mice. A sharp

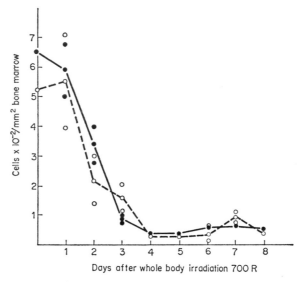

FIG. 7 Number of myelocytic cells in bone marrow of germ-free and conventional mice after whole body irradiation with a dose of 700 R. From Fliedner *et al.* (1967). ●—● =Conventional mice, ○ - - - ○ =germ-free mice.

decrease of the number of erythrocytic precursors is seen within 24 hr in both conventional and germ-free mice (Fig. 8) and minimal levels are reached at 3 days followed by a small "abortive" recovery, at least in the conventional animals. In the peripheral blood, the pattern of disappearance of leucocytes, erythrocytes and platelets was about the same in both the germ-free and conventional mice at this particular dose level, although the disappearance of platelets and granulocytes is somewhat slower in germ-free mice. This type of identical response would have been expected on the basis of the similar cell-kinetics of bone marrow cell systems in germ-free and conventional mice on the basis of the cytokinetic analyses performed by Bond *et al.* (1965) on the various cell renewal systems of the organism after whole body irradiation. Recently,

Boggs *et al.* (1967) reported a lower neutrophil blood concentration in germ-free mice (1120 ± 120 per mm³) than in conventional mice (1710 ± 140 per mm³). Their results of differential cell counts in the bone marrow and of neutrophil release following endotoxin injection or infection led to the conclusion that " . . . there is little to suggest that microorganisms or their products play a role in controlling neutrophil production in the normal steady state".

The data available thus far show only subtle differences in the kinetics of haemopoiesis between germ-free and conventional mice, which suggests that the absence of bacterial invasion remains the best explanation for the finding that germ-free animals are slightly more radioresistant as well as for their somewhat prolonged survival time after supralethal doses of irradiation.

IV. Cellular Kinetics of the Intestinal Epithelium and the Intestinal Syndrome in Germ-free Mice

The histological picture of the intestinal mucosa of germ-free animals shows a number of characteristic differences from that of conventional animals. The general aspect of the germ-free mucosa is usually described as "prenatal". The epithelial layer is more uniform in germ-free animals and the lamina propria is much less developed than in conventional animals. It contains only a few histiocytes and lymphocytes and an

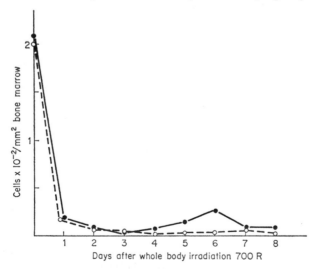

FIG. 8 Number of nucleated erythrocytic cells in bone marrow of germ-free and conventional mice after whole body irradiation with a dose of 700 R. From Fliedner *et al.* (1967). ● — ● =Conventional mice, ○ - - - ○ =germ-free mice.

occasional leucocyte in contrast to the physiological inflammation present in the conventional state. The patches of Peyer are smaller in the germ-free animal and show few reaction centres, little mitotic activity and a

Conventional   Germ–free

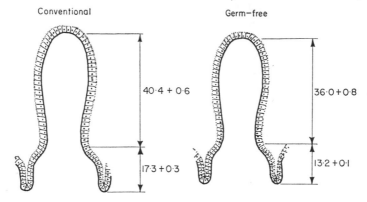

40·4 + 0·6   36·0+0·8

17·3 +0·3   13·2 +0·1

FIG. 9 Cytokinetics of ileal epithelium of conventional and germ-free mice. Data from Abrams *et al.* (1963).

No. of mitoses/crypt	$9·3\pm0·6$	$6·7\pm0·6$
No. of labelled cells/crypt	$11·0\pm0·4$	$7·6\pm0·6$
Rate of migration along villi in no. of cell positions/day	22	10

relatively small number of plasma cells. In germ-free guinea pigs the height of the villi was described as normal (Sprinz *et al.*, 1961). In germ-free mice the length of the villi seems to be somewhat less than in conventional animals, at least according to the number of epithelial cells counted. In both species the crypts of the germ-free animals are shallower. A thorough analysis of these differences in mice has been first presented by Abrams *et al.* (1963) who performed cell counts, quantitative estimates of cells labelled with tritiated thymidine and estimates of the rate of migration of the labelled cells along the villi.

These results, which are summarized in Fig. 9, showed that the total height of the mucosa is less in germ-free mice and that the ratio between the length of the villi and the length of the crypts is increased. More striking are the relatively decreased mitotic frequency in the crypts of germ-free mice and the decreased migration velocity of the cells along the surface of the villi, which leads to a life span of about double the value (4 days) found for epithelial cells in conventional mice (2 days).[1]

[1] Confirmation of these findings was provided by Lesher *et al.* (1964) for the mouse duodenum. The average transit time of the cells on the villus was doubled in the germ-free mouse. The size of the proliferative pool was decreased by about 37% and the generation rate of the proliferative cells was reduced by 25% so that the delivery of the cells by the crypts was decreased by a factor of 0·47.

On the basis of these data, Matsuzawa and Wilson (1965) have provided evidence that the survival time of animals exposed to radiation doses in the gastrointestinal death range is proportional to the life span of the intestinal cells on the villi, thereby offering an explanation for the increased survival time of germ-free animals.

At daily intervals after irradiation with a dose of 3000 R mice were sacrificed and the numbers of cells in the crypts and in the villi were counted in histological sections. In this series of experiments the mean

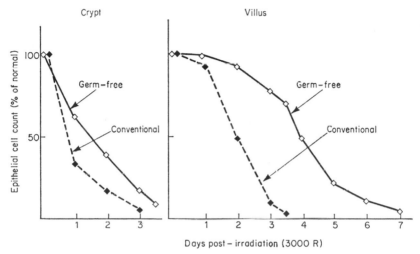

FIG. 10 Decrease of epithelial cell counts in crypts and villi following exposure to 3000 R of whole body irradiation in mice.

survival time of the conventional mice was 3·5 days (range 2·9–3·8 days) and of the germ-free mice 7·3 days (6·4–7·7 days). Crypts could no longer be identified after 3 days in the irradiated conventional mice and after 3·5 days in the irradiated germ-free mice. It can be seen from Fig. 10. that depletion of the crypts takes about 1 day longer in irradiated germ-free mice but that denudation of the villi to below 10% of normal cell numbers takes about twice as long in the germ-free animals as in the conventional mice. In both groups of irradiated animals death occurs roughly 0·5 day after complete denudation of the villi has been attained.

The same authors measured the rate of migration of the epithelial cells along the villi by administering a single dose of tritiated thymidine and studying autoradiographs of histological sections of the ileum at intervals in *non-irradiated* mice. In conventional animals the migration time of the labelled cells from the crypts to the top of the villi was 2·1 days; this interval was 4·3 days for the germ-free mice. The longer intervals that elapsed between irradiation and complete denudation of the villi presumably reflect the crypt-emptying times.

These results obtained with germ-free animals constitute an admirable confirmation of the cellular mechanism of the gastrointestinal syndrome. The experimental data do not seem to require an additional causative factor to explain the death of the animals at the times specified. Bacterial invasion which has been observed in mesenteric lymph nodes of mice irradiated with doses in excess of 2000 R (van der Waay, personal communication) does not seem to play a role in the development of fatal intestinal death. It may well be an important factor in determining the time of death for conventional animals in the supralethal dose region between the bone marrow syndrome and the survival time plateau of the intestinal syndrome, where cell production in the crypts has not been completely blocked. In these cases survival time may be adversely affected by the complication of septicaemia.

V. Responses of the Lymphatic System to Irradiation

Only one report dealing with this subject has come to the attention of the reviewer. Bauer *et al.* (1964) have studied the morphological changes in the lymphatic tissues of germ-free mice following a sublethal dose of whole body irradiation (see also Chapter 10). In addition, they measured the γ-globulin levels in the serum at intervals after irradiation.

Swiss Webster mice were exposed to 550 R of whole body irradiation which produced 35% mortality in the conventional animals and no deaths in the germ-free groups. The cellularity and the weight of the lymph nodes decreased sharply after irradiation in both groups and reached a minimum value of about 20% on the second day; thereafter gradual recovery occurred, which was still incomplete (about 75% of pre-irradiation values) on the fifteenth day after irradiation, this being the end of the observation period. At that time the recovery of the lymphatic tissues in the germ-free mice was slightly less but it is not clear whether this difference was statistically significant.

An increase of cell labelling with tritiated thymidine was observed in both groups, beginning on the first observation day (day 2) and continuing until the fifteenth day when the number of these cells was double the pre-irradiation values in both the germ-free and the conventional mice. The thymidine uptake was measured in all instances 1 hr after the injection.

An estimate was also made of the "immunologically competent cells" a term employed by the authors for "immunoblasts" and plasma cells as identified by light microscopy and for plasma cells which were found to contain γ-globulin by specific immuno-fluorescence. An increase of these cells was noted first on day 5 after irradiation and their number continued to rise until the tenth day in conventional animals and until the end of the observation period in the germ-free mice. Figure 11 contains the curves obtained, indicating that the rise was slower but nonetheless quite remarkable, in the germ-free mice. In contrast, serum γ-globulin levels

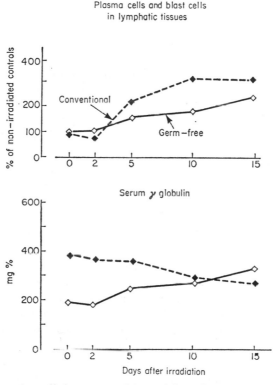

FIG. 11 Changes in cellular composition of lymphatic tissue and in serum γ-globulin levels following irradiation of mice with a whole body dose of 550 R. Data from Bauer *et al.* (1964).

decreased from 400 mg % to about 250 mg % on the fifteenth day in the conventional mice, but increased from 200 mg % to nearly 350 mg % in the germ-free animals.

As was to be expected the radiation-induced morphological damage to the lymphatic tissues seemed to be similar in both groups. Less easily explained are the striking differences in γ-globulin responses and the increase of the number of plasma cells in the germ-free mice. With regard to the first phenomenon, the authors favour the hypothesis that the irradiated conventional animals might be losing γ-globulin at an accelerated rate via the intestinal mucosa. Such leakage would be less or even absent in the germ-free animals because of a lesser degree of epithelial damage. Proof for this hypothesis is clearly needed and moreover these findings seem to be interesting enough to warrant a more detailed investigation.

An increase of cells containing γ-globulins and of plasma cells in general has been observed by others in conventional animals of a number of species and this has usually been interpreted as the reflection of an antibody response to invading microorganisms, that is, a reaction against infections. The finding of increased numbers of plasma cells in the irradiated germ-free mice does seem to rule out the causative role of living microorganisms in this reaction. Bauer et al. (1964) have pointed out two alternative explanations. The first implicates antigens derived from the food, the environment or from breakdown products of cells damaged by irradiation. Penetration of exogenous antigens could be facilitated in an irradiated organism. The inclusion of endogenous antigens in this hypothesis in fact amounts to the re-introduction of the concept of irradiation-induced autoimmune reactivity, which was repeatedly proposed before, notably by Allegretti (1960) and Allegretti et al. (1962). Germ-free animals clearly offer the possibility to investigate this concept more closely but it should be kept in mind that the germ-free animals thus far employed may harbour one or more viruses.

The second theory which was brought forward to explain the increase of plasma cells after irradiation is described as a non-specific stimulating effect of irradiation *per se*. This idea is based on several observations of an enhancement of antibody formation after whole body irradiation in rabbits, among others by Dixon and McConahey (1963) and by Taliaferro et al. (1964). It is generally felt that this enhanced antibody response — which occurs to some extent only when a certain time relationship between the irradiation and the antigen injection exists — is related to the distribution of lymphatic cells by the irradiation. The interpretation of the data of Bauer et al. (1964), in the light of the postulates mentioned above, does not seem likely in view of the lack of quantitative information in their paper (except for the curves reproduced in Fig. 11) and because it has not been convincingly shown that the rise in "immunologically competent" cells was distinct from a rise in lymphoblasts which normally occurs in sublethally irradiated animals.

VI. Other Responses in Irradiated Animals

Ledney and Wilson (1964) made a detailed study of the changes in body weight, water and food consumption which occur following whole body irradiation with a dose of 750 R. They compared these changes with those occurring after intraperitoneal administration of 10 μg of *E. coli* endotoxin, in order to test the hypothesis that the early decrease in water and food intake seen in conventional animals might be due to an increased absorption of endotoxins from the radiation-damaged intestines. The fact that no significant differences were found between germ-free and conventional animals with regard to food and water intake and loss of body weight during the first 24 hr after irradiation ruled out the involvement of endotoxin in these early reactions, in particular because the decreases were not less pronounced in the germ-free animals than in the conventional ones. The results obtained with these measurements at later intervals after irradiation do not support the supposed role of endotoxins either.

Serum-bound iron (SBI) levels were followed after whole body irradiation by Barry and Wilson (personal communication). In both conventional and germ-free mice SBI levels rose sharply from about 250 μg % to over 400 μg % during the first day after irradiation with a dose of 850 R. In the conventional animals this rise was followed by a gradual decrease to reach a minimum value of about 200 μg % between 4 and 6 days after irradiation. Thereafter the level slowly increased again. In the germ-free mice, however, the high level attained during the first 24 hr was maintained for the entire observation period of 11 days post irradiation. When the germ-free animals received 10 μg of *E. coli* or *Salmonella typhosa* endotoxin intraperitoneally at 24 hr after irradiation the SBI fell abruptly, so that the pattern resembled that of irradiated, conventional mice. The authors adopted the working hypothesis that the hypoferraemia of irradiated conventional animals is due to an increased liberation of endotoxin and they announce continuation of their investigations with animals monoassociated with microorganisms producing endotoxin and with others not producing endotoxin.

It should be pointed out however, that they did not test the effect of endotoxin injections on SBI in normal animals or in conventional irradiated animals, so that the available evidence at present is insufficient to allow any conclusions concerning the role of endotoxins in irradiation-influenced fluctuations of SBI. An additional difficulty is their finding of a different diurnal rhythm of SBI levels in germ-free and conventional mice. Interestingly, conventional mice showed a pronounced maximum at 6 p.m., while the maximum in germ-free mice occurred at 6 a.m., although the light-dark cycle was similar in both groups. It has not been reported whether these rhythms remained unchanged after irradiation, so that the changes observed might still have been caused by an alteration of the diurnal variations.

VII. Factors Modifying the Radiation Response in Germ-free Animals

A. THE EFFECT OF ASSOCIATION WITH SPECIFIC MICROORGANISMS

The germ-free animal provides one of the best tools for the study of the specific effects of different species of microorganisms on the course of the bone marrow syndrome. By contaminating germ-free animals with a single species of bacteria one can obtain information on the invasiveness of that species following irradiation and on the influence of that specific infection on survival time and the clinical course of the radiation disease. Surprisingly enough, relatively little effort has been made thus far to explore these aspects of the problem.

Wilson (personal communication) studied the effect of mono-association of germ-free mice with representatives of the intestinal flora on their responses following irradiation with a whole body dose of 850 R. Following 850 R which is 100% lethal both to conventional and germ-free mice, the former die between 5 and 13 days (50% mortality at 9 days) and the germ-free mice between 12 and 25 days (50% mortality at 14 days). The contaminations were performed 3 weeks before the irradiation to allow attainment of a state of adaptation to the microflora.

The survival curves of the animals associated with *Pseudomonas aeruginosa* were identical to those of the conventional animals; the latter were reported to harbour this microorganism. Mice carrying *Proteus vulgaris* as the sole microorganism died somewhat later than the conventional ones, but earlier than the germ-free animals. Association with *Streptococcus faecalis* or *Alcaligenes faecalis* did not alter the survival curves, but *E. coli* and *Clostridium difficile* contamination produced a marked increase of radio-resistance. Survival of 80% of the monoassociated animals was observed. These protective effects have not been explained and, as far as *E. coli* is concerned, such a protection was not observed by McLaughlin *et al.* (1964), who reported detailed results of irradiation of *E. coli*-associated mice in the dose region between 550 and 40,000 R. In general, their monoassociated animals behaved essentially similarly to their conventional animals. It will be of great interest to learn whether the striking protective effects observed at Lobund with certain associates will be extended in the future.

B. THE PROTECTIVE EFFECTS OF BACTERIAL ENDOTOXIN

The protective effects of bacterial endotoxins were discovered by W. W. Smith and co-workers (1957, 1958, 1966) who made an elaborate study of substances which increase the resistance of mice to infections in an attempt to provide protection against the lethal effects of irradiation.

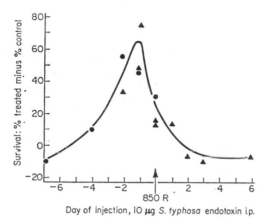

Day of injection, 10 μg *S. typhosa* endotoxin i.p.

FIG. 12 Effects of endotoxin administration on survival of lethally irradiated mice relative to the time of administration. Data from Smith *et al.* (1957). ● =AL/N mice, ▲ =CDBA mice.

They investigated the effects of endotoxins derived from three different Gram-negative bacteria in irradiated mice, rats and hamsters. The preparations were injected intraperitoneally either before or after irradiation. In many experiments a considerable decrease of mortality was obtained. In mice the results were in general more favourable when the endotoxin was injected before irradiation but in hamsters better results were obtained by post-irradiation administration. These beneficial effects were reproducible but it soon became evident that the time of injection of the endotoxin relative to the irradiation is very important, as illustrated in Fig. 12.

Initially Smith *et al.* (1958) thought that the protection was related to a decreased susceptibility to infection but later they showed that the endotoxins probably act by stimulating the haemopoietic regeneration. Subsequently it has been shown by Smith *et al.* (1966) that the injection of endotoxin induces a marked increase of the number of cells in the bone marrow which possess the capacity to produce colonies of haemopoietic cells in the spleen of lethally irradiated mice. These colony-forming cells are considered to represent the stem cells of the haemopoietic tissue and their stimulation could certainly account for an accelerated recovery of irradiated mice.

Ledney and Wilson (personal communication) have studied the protective effects of *E. coli* endotoxin in irradiated germ-free mice. The endotoxin was injected intraperitoneally 24 hr before irradiation, using a dose of 10 μg per mouse. In the radiation dose range which induces haemopoietic death an equal degree of protection was obtained in germ-free and conventional animals; dose reduction factors of 13% were found in both groups.

On the basis of measurements of tissue oxygen levels following endotoxin administration, Wilson *et al.* (1965) postulate that tissue hypoxia is responsible for the protective effects of endotoxin. The hypoxia was found to persist for several days after the administration of endotoxin. Pretreatment with endotoxin might therefore afford protection by two different mechanisms: one by way of the decreased radio-sensitivity of the tissues at the time of irradiation due to hypoxia, the second by way of a stimulative effect of hypoxia on haemopoiesis after the irradiation. The second mode of action is rather speculative however, in particular since Fried *et al.* (1966) have recently shown that a four-day period of hypoxia did not increase the content of colony-forming cells in the bone marrow of mice and Bruce and McCulloch (1964) even found a decrease of colony-forming cells after several days exposure to hypoxia.

C. BONE MARROW TRANSPLANTATION AS A TREATMENT OF LETHAL RADIATION INJURY

Lethal haemopoietic damage can be treated successfully in conventional animals by the intravenous administration of bone marrow cells. This form of treatment is quite effective in the lethal dose range representing the bone marrow syndrome but a beneficial effect is no longer observed after doses above 1200 R when lethal intestinal damage supervenes.

When the irradiated animals are protected from haemopoietic death with isologous bone marrow, the animals that survive for 30 days usually show an uneventful complete recovery. In contrast, treatment with homologous or heterologous bone marrow produces a similarly effective repopulation of the haemopoietic tissues and protection from acute death but a secondary disease develops in the 30-day survivors, killing a varying proportion of the radiation chimaeras in the course of the second and third month after irradiation. This secondary disease has been shown to be caused by an immunological reaction of the grafted cells against the tissues of the host (Van Bekkum and de Vries, 1967).

Experiments with germ-free mice have shed a new light on two aspects of protection with bone marrow transplantation: the first is concerned with the dose range in which the damage to the bone marrow is the critical factor determining survival, the second concerns the intricate pathogenesis of secondary disease.

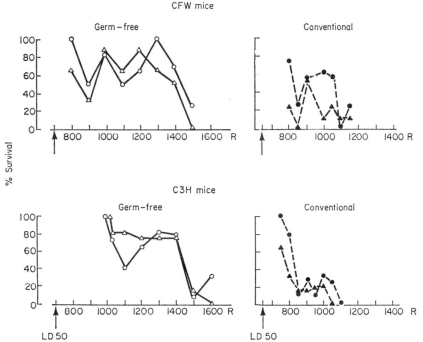

FIG. 13 Results of Connell and Wilson (1965) on bone marrow transplantation in lethally-irradiated mice. ● and ○, isologous bone marrow; ▲ and △, homologous bone marrow.

Connell and Wilson (1965) have treated lethally-irradiated germ-free mice with isologous and homologous bone marrow and compared the results with those obtained in conventional mice of the same strains receiving an identical treatment.

The mice were of the C_3H and CFW strains and received one femur equivalent (approximately 10^7 cells) of isologous marrow or two femur equivalents of homologous bone marrow after varying doses of whole body irradiation. The 30-day survival is depicted in Fig. 13. In both strains conventional mice responded slightly better to isologous marrow than to homologous marrow but the differences hardly appear to be significant.

The protection with isologous bone marrow in the animals employed is rather poor as compared to the results of other workers who obtain in conventional mice nearly 100% protection up to radiation doses of 1000 R with as little as 10^6 cells per mouse. It is possible that the strains employed have not been completely inbred but this probably could not wholly account for the defective protection. Far more likely is the explanation that the conventional mice were carriers of *Ps. aeruginosa* which has been reported to be the case in other publications from the same laboratory. It is well known that treatment with bone marrow yields variable and even erratic results in mice infected with *Pseudomonas* spp. and that in these cases protection is on the whole not very effective.

In the germ-free animals treated with bone marrow the survival is also quite variable. Obviously, this cannot be due to infection. The most remarkable difference with conventional mice is, however, that a substantial degree of protection was seen following doses as high as 1400 R, indicating that the damage to the intestinal tract does not become the determinant of mortality until the dose exceeds that value. In other words the bone marrow syndrome in germ-free mice seems to prevail in the dose range up to 1400 R.

The variable degree of protection with isologous bone marrow at lower irradiation doses is difficult to understand. Since the cause of death was not identified in these animals, only speculations can be provided. One, obvious to those experienced with germ-free techniques, is that some of the bone marrow recipients may not have received the full amount of bone marrow cells. The volume injected was only 0·1 ml; such a small volume may not have reached the circulation in some cases since the injections were made into the heart, which technique carries a large risk of missing the cardiac lumen. In view of these considerations, it does not seem justified to attach great significance to the differences in the responses to isologous and homologous bone marrow in the germ-free mice.

In a later report the authors have stated that pathological changes characteristic of secondary disease were not seen in germ-free mice treated with homologous bone marrow (Wilson and Connell, 1964). This statement carries important consequences with regard to the pathogenesis of secondary disease. The most characteristic lesions of secondary disease are found in the intestinal tract, liver, skin and lymphatic tissues and these have been described in many different species, including monkeys and man (van Bekkum *et al.*, 1966).

The lesions consist of cell necrosis and inflammatory reactions in the epithelial tissues and extreme atrophy of the lymphatic organs. The latter condition is generally implicated as being the cause of the increased susceptibility to infections during secondary disease. The course of events leading to the characteristic organ lesions of secondary disease has been postulated in two different ways.

In the first, the lymphatic atrophy is considered the primary event, resulting in turn from the reaction of the grafted lymphoid cells against the host (Loutit and Micklem, 1962).

graft *vs* host reaction $\longrightarrow$ lymphatic atrophy $\longrightarrow$ infections and specific lesions

Certain lesions were considered to be the direct consequence of the infectious processes, others were ascribed to a lack of "trophic" function of lymphocytes.

The second hypothesis attributes the organ lesions directly to the action of the grafted lymphoid cells upon the host epithelia.

Graft *vs* host reaction
- lymphatic atrophy→infections
- epithelial necrosis complicated by infections

The use of germ-free animals seemed to provide the possibility to distinguish between the two mechanisms. Accordingly, a microscopical investigation was made of the tissues of the lethally-irradiated germ-free mice which were transplanted with rat bone marrow (van Bekkum *et al.*, 1966). Mice were examined at intervals up to 80 days after irradiation, when it could be shown that they still were complete chimaeras.

In the lymphatic tissues, skin, liver and intestinal tract, lesions were found that were characteristic of graft *vs* host reaction. In contrast to the findings in conventional animals these lesions were not complicated by inflammatory changes and thus provided a pure picture of graft *vs* host reactions in the various tissues. These results have finally confirmed the secondary nature of infections in the pathology of secondary disease and moreover the use of germ-free animals has permitted the observation of uncomplicated graft *vs* host reactions.

VIII. Conclusions

The most important contribution of gnotobiotics to radiation biology so far has been in increasing the availability of animals free of endemic infections for experiments involving high doses of whole body irradiation. The possibilities of eliminating interfering microorganisms from laboratory animal colonies have been facilitated enormously by the use of germ-free or ex-germ-free foster mothers to raise surgically derived young. The benefits of this change from conventional to disease-free animals have now become clearly recognizable from results of short term experiments. For long term experiments such profits are still to be collected but there is little doubt that the results will be equally rewarding. On the other hand, the number of specific radiobiological problems that have been elucidated with the use of germ-free techniques is yet limited. The reason is no doubt that the germ-free technique is still so complicated and cumbersome that its use is restricted to relatively few laboratories. It is not likely that the introduction of cheap prefabricated isolators, sterilized food, etc. will improve this situation substantially because, not only are optimal tools required, but also skill and especially experience with the extensive precautions and bacteriological controls that remain necessary to produce valid results with germ-free animals.

It is not surprising therefore that the majority of the papers dealing with radiobiological research in germ-free animals have emanated from Lobund Institute. The fact that several radiobiological research centres have recently installed gnotobiotic facilities justifies the hope that much more and diverse information concerned with the role of micro-organisms in radiation effects of the mammalian organism will become available in the near future.

Yet a few important problems have been clarified by the use of germ-free animals, notably the role of the microflora in the symptomatology and pathogenesis of the gastrointestinal syndrome and the contribution of infectious processes to the development of the specific lesions of secondary disease following foreign bone marrow transplantation in lethally-irradiated animals. Several detailed studies of haemopoiesis in normal and irradiated germ-free animals can be expected to appear in the literature soon. There seems to be little doubt that the results so far obtained will stimulate further investigations of the effects of specific microorganisms on the development and manifestations of the various forms of radiation injury.

REFERENCES

Abrams, G. D., Bauer, H. and Sprinz, H. (1963). *Lab. Invest.* **12**, 355–364.

Allegretti, N. (1960). *Bull. Sci.* **5**, 77.

Allegretti, N., Vitale, B. and Dekaris, D. (1962). *Int. J. Radiat. Biol.* **4**, 363–370.

Bauer, H., Horowitz, R. E., Paronetto, F., Einheber, A., Abrams, G. D. and Popper, H. (1964). *Lab. Invest.* **13**, 381–388.

van Bekkum, D. W. and de Vries, M. J. (1967). "Radiation Chimaeras", Logos Press Ltd. and Academic Press, London and New York.

van Bekkum, D. W., de Vries, M. J. and van der Waay, D. (1966). *J. natn. Cancer Inst.* **38**, 223–231.

Boggs, D. R., Chervenick, P. A., Marsh, J. C., Pilgrim, H. I. and Cartwright, G. E. (1967). *59th A. Mtg. Am. Soc. clin. Invest.*, **59**, 19.

Bond, V. P., Fliedner, T. M. and Archambeau, J. O. (1965). "Mammalian Radiation Lethality", Academic Press, New York.

Bruce, W. R. and McCulloch, E. A. (1964). *Blood* **23**, 216–232.

Conard, R. A., Cronkite, E. P., Brecher, G. and Strome, C. P. A. (1956). *J. appl. Physiol.* **9**, 227–233.

Connell, J. and Wilson, R. (1965). *Life Sciences* **4**, 721–729.

Curran, P. F., Webster, E. W. and Hovsepian, J. A. (1960). *Radiat. Res.* **13**, 369–380.

Dixon, F. J. and McConahey, P. J. (1963). *J. exp. Med.* **117**, 833–848.

Fliedner, T. M., Fache, I. and Adolphi, C. (1966). *Schweiz. med. Wschr.* **96**, 1236–1238.

Fliedner, T. M., Stehle, H., Teitge, H. and Fache, I. (1967). To be published.

Fried, W., Martinson, D., Weisman, M. and Gurney, C. W. (1966). *Expl Haemat.* No. **10**, 22–24.

Ledney, G. D. and Wilson, R. (1964). *Radiat. Res.* **22**, 207–208.

Lesher, S., Walburg, H. E. and Sacher, G. A. (1964). *Nature, Lond.* **202**, 884–886.

Loutit, J. F. and Micklem, H. S. (1962). *Br. J. exp. Path.* **43**, 77–87.

Matsuzawa, T. (1965). *Tohoku J. exp. Med.* **85**, 257–263.

Matsuzawa, T. and Wilson, R. (1965). *Radiat. Res.* **25**, 15–24.

McLaughlin, M. M., Dacquisto, M. P., Jacobus, D. P., Forbes, M. and Parks, P. E. (1958). *Radiat. Res.* **9**, 147.

McLaughlin, M. M., Dacquisto, M. P., Jacobus, D. P. and Horowitz, R. E. (1964). *Radiat. Res.* **23**, 333–349.

Quastler, H. (1956). *Radiat. Res.* **4**, 309–320.

Quastler, H. and Sherman, F. G. (1959). *Expl Cell Res.* **17**, 420–438.

Reyniers, J. A., Trexler, P. C., Scruggs, W., Wagner, M. and Gordon, H. A. (1956). *Radiat. Res.* **5**, 591.

Sherman, F. G. and Quastler, H. (1960). *Expl Cell Res.* **19**, 343–360.

Smith, J. C. (1960). *A.M.A. Archs Path.* **70**, 94–102.

Smith, W. W., Alderman, I. M. and Gillespie, R. E. (1957). *Am. J. Physiol.* **191**, 124–130.

Smith, W. W., Alderman, I. M. and Gillespie, R. E. (1958). *Am. J. Physiol.* **192**, 263–267.

Smith, W. W., Brecher, G., Budd, R. A. and Fred, S. (1966). *Radiat. Res.* **27**, 369–374.

Sprinz, H., Kundel, D. W., Dammin, G. J., Horowitz, R. E., Schneider, II. and Formal, S. B. (1961). *Am. J. Path.* **39**, 681–695.

Sullivan, M. F., Hulse, E. V. and Mole, R. H. (1965). *Br. J. exp. Path.* **46**, 235–244.

Swift, M. N. and Taketa, S. T. (1956). *Am. J. Physiol.* **185**, 85–91.

Taliaferro, W. H., Taliaferro, L. G. and Jaroslow, B. N. (1964). "Radiation and immune mechanisms", Academic Press, New York.

Teitge, H. (1967). Inaugural-Dissertation der Medizinischen Fakultät Freiburg.

Walburg, H. E., Mynatt, E. I. and Robie, D. M. (1966). *Radiat. Res.* **27**, 616–629.

Wensinck, F., van Bekkum, D. W. and Renaud, H. (1957). *Radiat. Res.* **7**, 491–499.

Wilson, R. (1963). *Radiat. Res.* **20**, 477–483.

Wilson, R. and Connell, J. (1964). *Expl Haemat.* No. **7**, 56–57.

Wilson, R. and Piacsek, B. (1962). *Fedn Proc. Fedn Am. Socs exp. Biol.* **21**, 423.

Wilson, R., Matsuzawa, T. and Connell, J. (1964). *Radiat. Res.* **22**, 249–250.

Wilson, R., Ledney, G. D. and Matsuzawa, T. (1965). *Prog. Biochem. Pharmacol.* **1**, 622–626.

Author Index

The numbers in *italics* refer to the pages on which the references are listed

Subject Index

A

acetylcholine, caecal wall content of, rat, 132
  caecal wall sensitivity to, rat, 132
  metabolic effects, rat, 140
Actinobacillus, hydrolysis of urea, caecum, rat, 165
adenine, dietary, effect on toxicity of orotic acid, rat, 174
adrenal glands, effect of overcrowding, rat, 53
  size, rat, 122
adrenaline, metabolic effects, rat, 140
L-adrenaline, sensitivity of caecal wall to, 132
Aerobacter, intestinal, effect, folic acid deficiency, rat, 167
age effects, 147
agglutinins, rat, chicken, 202
air
  humidity, isolators, 8
  movement, isolators, 52
  pressure, prevention of contamination, isolators, 6
  sterilization, 8
  temperature, isolators, 8
albumin, *see* bovine serum albumin
Alcaligenes faecalis, effect on survival time, irradiated mouse, 260
Alcaligenes, intestinal, effect, folic acid deficiency, rat, 167
alfalfa, inhibition of *Bacteroides melaninogenicus*, mouse, 39
allodeoxycholic acid metabolism, 191
alopecia, kitten, 107
aluminium, effect of sterilizing agents, 6
amino acids, *see also* cysteine, histidine, lysine, methionine, threonine, tyrosine
  caecal content, rat, 137, 164
ammonia, caecal content, rat, 164
amylase(s), *see also* enzymes, digestive

excretion, rat, 138
  intestinal content, chicken, 138, 163
    rat, 163
anaemia, irradiated mouse, 249
anaemia, folic acid deficiency, rat, 167
animal production, large mammals, 64
  small mammals, 47
  economics, 51
anomalies, 127
antibiotics, dietary, effect, pantothenic acid deficiency, rat, 168
antibiotics, dietary, effect, growth chicken, 172
antibodies, natural, 219
  production, 198, 203, 211
antigens, contaminating organisms, effect on immunological experiments, 40
  response to, 217
anus, sucking, guinea pig, 54
apoferritin, *see* ferritin
ascorbic acid, dietary, effect, pantothenic acid deficiency, rat, 168
  effect of food sterilization, 91
  requirement, 93
    effect of intestinal microflora, guinea pig, 168
Aspergillus niger, effect on intestinal bile acids, rat, 189
atropine, metabolic effects, rat, 140
autoclave, effect on temperature in isolator, 52
autoclaving, *see* sterilization, steam
auto-immune reaction, irradiated mouse, 258
autoimmune stimuli, 220

B

Bacillus subtilis contamination, chicken, 39
bacteria, *see* microorganisms
Bacteroides melaninogenicus inhibition by alfalfa, 39